WITHDRAWN

UMIST Library and Information Service

This book should be returned on or before
the last date below unless recalled earlier.
Overdue items will be liable to a fine

Umist Library and Information Service
P.O. Box 88, Manchester, M60 1QD

Design Analysis

An integral part of engineering design is the development of models that describe physical behavior or phenomena in mathematical terms. As engineering systems have become more complex, classic linear methods of modeling and analysis have proved inadequate, creating a need for nonlinear models to solve design problems.

This text provides an introduction to mathematical modeling of linear and nonlinear systems, with an emphasis on the solution of nonlinear design problems. While encouraging the use of the computer as a tool for modeling and analysis, the aim is to discuss the basic concepts underlying computer techniques and to seek analytical solutions. Among topics covered are exact solution, numerical solution, graphical solution, and approximate solution methods; and the stability of nonlinear systems. Numerous examples throughout the book demonstrate the application of specific modeling methods to a wide variety of real engineering systems. End-of-chapter problems offer an added resource in learning to model and analyze nonlinear systems. Among case studies included in the final chapter are a light bulb model, automobile carburetor model, surge analysis for a hydroelectric plant, and fluid mechanics of blood flow in the kidney.

Intended for senior undergraduate or beginning graduate students, this text also serves as a helpful reference for engineers who can put into practice the mathematical modeling methods and solution techniques presented here.

Professor David E. Thompson is Chairman of the Mechanical Engineering Department at the University of New Mexico.

Design Analysis

Mathematical Modeling of Nonlinear Systems

David E. Thompson

PUBLISHED BY THE PRESS SYNDICATE OF THE UNIVERSITY OF CAMBRIDGE
The Pitt Building, Trumpington Street, Cambridge, CB2 1RP, United Kingdom

CAMBRIDGE UNIVERSITY PRESS
The Edinburgh Building, Cambridge CB2 2RU, UK   http://www.cup.cam.ac.uk
40 West 20th Street, New York, NY 10011-4211, USA   http://www.cup.org
10 Stamford Road, Oakleigh, Melbourne 3166, Australia

© Cambridge University Press 1999

This book is in copyright. Subject to statutory exception
and to the provisions of relevant collective licensing agreements,
no reproduction of any part may take place without
the written permission of Cambridge University Press.

First published 1999

Printed in the United States of America

Typeset in Times Roman 10.5/13 pt. and Univers in LATEX 2_ε [TB]

A catalog record for this book is available from the British Library

Library of Congress Cataloging-in-Publication Data
Thompson, David E., 1940–
 Design analysis : mathematical modeling of nonlinear systems /
David E. Thompson.
  p.   cm.
  ISBN 0-521-62170-4 (hb)
  1. Engineering design—Mathematical models. 2. Nonlinear systems—
Mathematical models.
 TA174.T55   1999
 620′.0042′01511B—dc21   98-3588
                                                                    CIP

ISBN 0 521 62170 4 hardback

Contents

Preface		*page* xi
List of Figures		xiii
List of Tables		xvii
Acknowledgments		xix

1 The Design Process — 1
  1.1 Elements of the Design Process — 2
    1.1.1 Identification — 4
    1.1.2 Defining the Problem — 4
    1.1.3 Ideation — 5
    1.1.4 Selection — 5
    1.1.5 Analysis — 6
    1.1.6 Implementation — 7
    1.1.7 Evaluation — 7
  1.2 The Computer as a Design Tool — 7
  1.3 Summary — 8

2 Mathematical Modeling — 10
  2.1 Special Behavior of Nonlinear Systems — 13
  2.2 Modeling Basics — 14
    2.2.1 Purpose — 16
    2.2.2 Resources — 16
    2.2.3 Expert Knowledge — 16
    2.2.4 Model Validation — 18
    2.2.5 Constraints — 18
    2.2.6 Examples of Modeling — 18
    2.2.7 Method of Lagrange's Equations — 22
    2.2.8 Vector Notation and Operations — 25
      2.2.8.1 Elementary Operations — 25
      2.2.8.2 $\vec{\nabla}$-Operators — 25
      2.2.8.3 Divergence Theorem — 26
      2.2.8.4 Stokes' Theorem — 26
      2.2.8.5 Fundamental Vector Properties — 27
      2.2.8.6 Introduction to Tensor Notation — 29
      2.2.8.7 Tensor Algebra — 30
      2.2.8.8 Properties of Second-Order Tensors — 30

2.2.8.9	The Alternating Tensor	32
2.2.8.10	Derivatives of Tensors	33
2.2.8.11	Isotropic Tensors	33
2.2.8.12	Vector-Tensor Equivalence	34

- 2.3 Integral Equations — 34
  - 2.3.1 Basic Laws for Continuous Matter — 34
  - 2.3.2 General Conservation Formulation — 35
    - 2.3.2.1 Conservation of Matter — 38
    - 2.3.2.2 Conservation of Momentum — 39
  - 2.3.3 First Law of Thermodynamics — 40
- 2.4 Differential Equations — 41
  - 2.4.1 Ordinary Differential Equations — 42
    - 2.4.1.1 Linear Differential Equations — 42
    - 2.4.1.2 Nonlinear Differential Equations — 42
  - 2.4.2 Partial Differential Equations — 43
- 2.5 Examples of Nonlinear Systems — 43
  - 2.5.1 Heat Transfer — 43
  - 2.5.2 Thermodynamics — 43
  - 2.5.3 Mechanics — 44
  - 2.5.4 Simple Suspension System — 44
  - 2.5.5 Fluid Mechanics — 45
  - 2.5.6 Continuity Equation — 45
  - 2.5.7 Blasius Equation — 45
  - 2.5.8 Vehicle Dynamics — 46
- 2.6 Normalization of Equations — 47
  - 2.6.1 Complex Nonlinear Problems — 59
  - 2.6.2 Infinite and Semi-Infinite Domains — 59
- 2.7 Transformation of Variables — 60
  - 2.7.1 Exponential Transformations — 61
  - 2.7.2 Generalized Transformation — 62
  - 2.7.3 Similarity Transformations — 63
  - 2.7.4 Well-Posed Problems — 65
  - 2.7.5 Problems with Missing Boundary Conditions — 67
  - 2.7.6 Summary — 68
- 2.8 Understanding Differential Equations — 68
  - 2.8.1 Force on a Spring — 69
  - 2.8.2 Force on a Damper — 69
  - 2.8.3 Force on a Mass — 69
  - 2.8.4 Force on a Spring-Damper Pair — 69
  - 2.8.5 Force on a Spring-Mass Pair — 70
  - 2.8.6 Force on a Mass-Dashpot Pair — 70
  - 2.8.7 Force on a Spring-Mass-Dashpot System — 71
  - 2.8.8 Other Forms of Nonlinear Equations — 71
- 2.9 Comparison of Model and Experiment — 72
  - 2.9.1 Concepts of a "Best" Model — 72
    - 2.9.1.1 Error Measures — 73

	2.9.1.2 Minimum (-Maximum)	74
	2.9.1.3 Least Sum Square Error	75
2.9.2	Other Natural Error Measures	77
2.9.3	Nonlinear Problems	77
	2.9.3.1 Gradient Method	78
	2.9.3.2 Nongradient Method	78
2.9.4	Constrained Problems	80
	2.9.4.1 Variable Mapping	81
	2.9.4.2 Penalty Functions	81
2.10	Summary	83
2.11	Problems	84

3 Exact Solution Methods — 87

- 3.1 Technique of Linearization — 87
  - 3.1.1 Method of Finding Linear Coefficients — 88
  - 3.1.2 Linearization of a Multivariable Function — 90
  - 3.1.3 Taylor-Series Expansion Errors — 94
    - 3.1.3.1 Classical — 94
    - 3.1.3.2 First-Term-Neglected Method — 95
- 3.2 Direct Integration — 96
  - 3.2.1 Radiatively Heated Thermal Capacitance — 96
  - 3.2.2 Spring-Dashpot System — 98
  - 3.2.3 Automobile Handling Revisited — 99
  - 3.2.4 Gravitational Attraction — 100
- 3.3 Variation of Parameters — 101
- 3.4 Equations Leading to Elliptic Integrals — 103
  - 3.4.1 Suspended Mass — 104
  - 3.4.2 Simple Pendulum — 106
- 3.5 Power-Series Method — 108
  - 3.5.1 Spring-Damper Problem — 108
- 3.6 Picard's Method — 109
- 3.7 Reversion of Power Series — 112
- 3.8 Summary Comments — 113
- 3.9 Problems — 114

4 Numerical Solution Methods — 118

- 4.1 Taylor-Series Method — 118
- 4.2 Euler Method — 120
  - 4.2.1 Modified Euler Method — 121
  - 4.2.2 Extension of Euler Method to Higher-Order Systems — 122
- 4.3 Runge-Kutta Method — 123
  - 4.3.1 Solution of Two Simultaneous Equations — 126
  - 4.3.2 Final Remarks on Runge-Kutta Method — 127
- 4.4 Multistep Methods — 128
- 4.5 Step-Size Determination — 129
- 4.6 Summary Comments — 130
- 4.7 Problems — 130

5 Graphical Solution Methods — 133
- 5.1 First-Order Equations (Method of Isoclines) — 133
  - 5.1.1 Phase-Plane Analysis — 135
  - 5.1.2 Recovery of the Independent Variable — 136
- 5.2 Second-Order Systems — 137
  - 5.2.1 Linear Spring-Mass System — 137
  - 5.2.2 Method of Isoclines — 138
  - 5.2.3 Effect of Normalization — 138
- 5.3 Liénard's Method — 139
- 5.4 Pell's Method — 142
- 5.5 Summary Comments — 144
- 5.6 Problems — 144

6 Approximate Solution Methods — 148
- 6.1 Method of Perturbation — 148
  - 6.1.1 Intended Area of Applicability — 149
  - 6.1.2 Solution Technique — 149
- 6.2 Iteration Technique — 155
- 6.3 Power-Series Method — 157
  - 6.3.1 Duffing Equation — 159
  - 6.3.2 Generalized Duffing Equation — 161
- 6.4 Method of Harmonic Balance — 163
  - 6.4.1 Equivalent Linear Equation — 166
- 6.5 Galerkin's Method — 168
  - 6.5.1 Secondary Boundary Conditions — 171
- 6.6 Summary Comments — 180
- 6.7 Problems — 181

7 Stability of Nonlinear Systems — 184
- 7.1 Routh Method — 185
- 7.2 Singular-Point Analysis — 187
  - 7.2.1 General System — 188
  - 7.2.2 Linear Transformations — 189
  - 7.2.3 Classification of Singularities — 192
    - 7.2.3.1 Real Roots, Same Sign — 193
    - 7.2.3.2 Real Roots, Different Sign — 194
    - 7.2.3.3 Complex Roots (Pure Imaginary) — 194
    - 7.2.3.4 Complex Conjugate Roots — 195
  - 7.2.4 Examples of Singularity Analysis — 196
- 7.3 Poincaré Index — 202
- 7.4 Bendixson's First Theorem — 206
- 7.5 Second Method of Lyapunov — 210
  - 7.5.1 Lyapunov Function — 211
  - 7.5.2 First Theorem of Lyapunov — 212
  - 7.5.3 Second Theorem of Lyapunov — 215

	7.6 Introduction to Chaotic Systems	217
	7.6.1 Phase Space	218
	7.6.2 Poincaré Sections	219
	7.6.3 Bifurcation Diagrams	220
	7.7 Summary Comments	220
	7.8 Problems	221
8	**Case Studies**	**226**
	8.1 Lightbulb Model	226
	8.1.1 Method 1: Linearization	226
	8.1.2 Method 2: Exact Solution of Nonlinear Problem	228
	8.1.3 Method 3: Galerkin Solution	228
	8.1.4 Method 4: Numerical Solution	229
	8.1.5 Summary Comments	229
	8.2 Automobile Carburetor Model	229
	8.3 Automobile Engine Model	231
	8.3.1 Model Development	232
	8.3.2 Experimental Data	233
	8.4 Surge Analysis for a Hydroelectric Power Plant	233
	8.4.1 Development of the Model	234
	8.4.2 Numerical Integration Scheme	235
	8.4.3 Results	235
	8.5 Constant-Deceleration Shock Absorber	236
	8.5.1 Equations of Motion for Incompressible Flow	238
	8.5.2 Procedure for Constant-Deceleration Design	241
	8.6 Fluid Mechanics of Blood Flow in the Kidney	242
	8.6.1 Exact Solution	244
	8.6.2 Galerkin Solution	244
	8.6.3 Comments	245
	8.7 Torque and Motion of the Finger	245
	8.8 Summary Comments	246
	8.9 Problems	247
References		267
Index		271

Preface

The intent of this text is to give upper-division undergraduate and graduate students a new and more general analytical approach to design while continuing to emphasize the use of the computer as a modeling and analysis tool. This material builds on the modeling training that all undergraduates receive (albeit in an ad hoc manner), and places students at ease in attempting to model systems on their own. Students are encouraged to exercise their creative and imaginative talents in building comprehensive models of real physical systems. All of my undergraduate and graduate students can, for example, readily find the points of maximum stress in a mechanical assembly. Few would attempt to express the design in a manner that would allow them to minimize the cost of the assembly, determine the relationship between the cost and the power requirements, the rotating inertia of the assembly, or other entities, whether "liabilities" or "assets." After instruction in the concepts of optimal design and modeling fundamentals expressed in this text, even when their systems became complex and nonlinear, these same students then were able to extend the basic models they learned in engineering and to proceed to several forms of both analytic and numerical solutions.

In considering my own training, I found that there were identifiable places and times at which I learned each of my design skills:

Basic engineering models—undergraduate engineering courses.
Measurement of actual systems—work as a consultant and at General Dynamics' Fort Worth division and my research studies at Louisiana Tech and Indiana University's School of Medicine.
Developing my own models—principally through my own efforts during my doctoral research.
Solution techniques for nonlinear systems—nonlinear methods course taught by Professor Robert Kohr at Purdue.
Comprehensive model development—teaching design and in my research while at Louisiana State University.

Thus, it seemed that my own experience was perhaps unusual, and that my modeling and analysis skills resulted from a serendipitous assortment of courses, research problems, and experiences, and from my own natural curiosity. It seemed appropriate, therefore, to place into the literature a text that organizes these various basic modeling concepts so that the path to good design and analysis, now frequently lumped together and termed simulation, would be more formalized and less haphazard.

A great deal of emphasis in this text has been placed on the analytic approach to the solution of nonlinear models. As our world becomes more complex, I anticipate that nonlinear problems will become more common. Certainly, as we approach the limits of our knowledge of the behavior of materials and composite structures, and as we seek to reduce weight and improve performance, it is expected that nonlinear behavior will become commonplace. In such designs, those engineers having the proper modeling and nonlinear analysis skills will become highly valuable to their employers.

This text has been taught at both Louisiana State University and at the University of New Mexico as an advanced senior elective or beginning graduate-level course for many years. The scope of the material has evolved over time, however, to the point where I now find it impossible to cover all of the topics in the text during the course of a semester. Since some of the material is covered in other courses (e.g., the numerical analysis subjects covered in Chapter 4), this is not a severe deficiency. It is also hoped that this text will serve as a reference for engineers as they begin to practice the methods and techniques presented.

David E. Thompson

List of Figures

1.1	Engineering design as (a) an iterative process and (b) a functional mapping.	2
1.2	Levels of abstraction in engineering design.	3
2.1	Schematic representation of mathematical modeling of real problems.	11
2.2	Stress-strain behavior of plain carbon steel.	11
2.3	Comparison of Newton formulation (solid line) and polynomial (dashed line) representations.	12
2.4	Response of a linear system (principle of superposition).	13
2.5	Comparison of input and output for linear and nonlinear systems.	13
2.6	Application of Occam's razor to modeling.	15
2.7	Spring-driven toy car.	17
2.8	Improved model for stacking eggs.	19
2.9	Sketch of a whale and some related information.	21
2.10	Outward area normal vector definition.	26
2.11	Simple rotation transformation.	27
2.12	A comparison of system and control volume boundaries.	36
2.13	System, and control volume at two infinitesimally separate times.	36
2.14	Simple rocket car on wheels.	39
2.15	Simple pendulum in a gravitational field.	44
2.16	Simplified model of automotive steering problem.	46
2.17	Effect of steering angle and yaw velocity on slip angle.	47
2.18	Response of a normalized spring-mass-damping system with varying damping.	51
2.19	Schematic of a temperature-dependent RC network.	51
2.20	Normalized first-order response ($R =$ constant).	52
2.21	Coupled normalized current and temperature response of a temperature-dependent resistance.	55
2.22	Diagram of the balloon-catheter system to be analyzed.	56
2.23	Pipe draining a reservoir.	58
2.24	Normalized reservoir-pipe exit velocity versus time.	59
2.25	The effect of varying the model parameter on the model performance.	74
2.26	Several search cycles for Powell's method.	80
2.27	Cost penalty function.	82
2.28	A circular arc of angle θ.	85
3.1	Function linearization over the domain $x[A, B]$.	88
3.2	Simple flow regulator system.	90
3.3	Flyball governor system.	91
3.4	Exact, linearized solutions for a radiatively heated thermal capacitance.	98

3.5	Unforced nonlinear system with square-law spring, linear dashpot.	98
3.6	Simple gravitational attraction problem.	100
3.7	Suspended-mass problem.	104
3.8	Linearized versus exact solution obtained using elliptic integrals.	106
3.9	Problem domain for Picard's method.	110
3.10	A mass suspended on a wire.	114
3.11	Simple switched diode-capacitor circuit.	116
4.1	Solution to nonlinear system problem.	119
4.2	Cumulative numerical errors arising from Euler's method.	121
4.3	Modified Euler method applied using forward and backward differences.	122
4.4	Comparison of Euler and modified Euler methods.	122
4.5	Graphical representation of the Runge-Kutta method.	124
5.1	Method of isoclines.	134
5.2	Isocline solution for a nonlinear spring-dashpot system.	135
5.3	Phase-plane solution of $dx/dt = f(x)$.	135
5.4	Solution of a massless nonlinear spring problem.	136
5.5	Phase-plane solution for the nonlinear spring-dashpot system.	137
5.6	Isocline phase-plane solution for a harmonic system.	138
5.7	Nonnormalized phase-plane solution.	139
5.8	Example of a single-valued function.	140
5.9	Function $x + f_1(v) = 0$ in the phase plane.	140
5.10	Solution of van der Pol oscillator.	141
5.11	Simple brake system.	141
5.12	Typical frictional forces in a simple brake.	142
5.13	Liénard's solution for a simple braking problem.	142
5.14	Pell's method.	143
5.15	Phase-plane solution for a simple pendulum.	144
5.16	Bistate force generated by a frictional element acting on a rotating member with relative velocity $(v - r\omega)$.	145
5.17	Bang-bang servo control of a ship's heading.	146
5.18	Bistate switching moment generated by a bang-bang servomechanism controlling a ship's rudder.	146
6.1	Perturbation solution to the automobile handling problem.	151
6.2	Jump resonance phenomenon of a nonlinear system.	160
6.3	Simplified model of an automobile suspension system.	160
6.4	Force-compression of an automobile tire.	161
6.5	Jump resonance of an automobile suspension system.	161
6.6	Generalized Duffing system.	161
6.7	Frequency content of a Duffing system: (1) subharmonics exist below this frequency; (2) this frequency, only the subharmonic exists; (3) A_1 for a linear system.	163
6.8	Comparison of Galerkin and exact solutions for a spring-dashpot system.	173
6.9	Buckling of a simple column under axial loading.	176
6.10	Buckling modes of an axially loaded column.	176
6.11	Temperature variations in an infinite wall.	178
6.12	Two masses suspended on linear springs and connected by a cubic spring.	180

List of Figures

6.13	Galerkin versus exact solutions for two masses between nonlinear springs.	180
6.14	Functional relation of $f(x)$ to x.	182
7.1	Transformation of origin to singular point.	188
7.2	Stretching, compression, rotation from real-valued, linear transformations.	192
7.3	Graphical types yielding nodes.	194
7.4	Real roots with different signs produce SADDLEs.	194
7.5	Complex pure imaginary roots produce CENTERs.	195
7.6	Definition of transformation.	196
7.7	Complex conjugate roots produce a FOCUS.	196
7.8	Singular-point stability analysis for a mass on a fifth-order spring.	199
7.9	Functional $f(k, v) = e^{-kv}v^k$.	202
7.10	Relationship between predator and prey populations.	202
7.11	Closed path around a singularity.	203
7.12	Poincaré index of singularities.	203
7.13	Phase-plane solution for a van der Pol oscillator.	207
7.14	Schematic of generator and its operational behavior.	207
7.15	Stability graph for the generator.	209
7.16	Simple linear controller.	215
7.17	Nonlinear controller.	216
7.18	Transient response and limit cycle of a damped, forced pendulum: (a) Pendulum angle versus time, $q = 2$, $g = 1.5$, $\theta_0 = 2$, $\omega_0 = 0$; (b) Phase plot, pendulum velocity versus angle.	218
7.19	Chaotic response of a damped, forced pendulum: (a) Pendulum angle versus time, $q = 2$, $g = 3.1$; (b) Phase plot, pendulum velocity versus angle.	218
7.20	Trajectory cross-sectional areas.	219
7.21	Poincaré section for a damped, forced pendulum.	220
7.22	Bifurcation diagram for the sensitivity of a damped pendulum to driving amplitude, g.	220
7.23	Schematic of a manually operated garage door.	224
8.1	Simplified lightbulb model.	227
8.2	Comparison of the linearized, exact, and Runge-Kutta solutions.	227
8.3	Simplified diagram of an automobile carburetor.	229
8.4	Normalized fuel flow at varying carburetor throttle angles.	231
8.5	Simplified diagram of an automobile engine.	232
8.6	Diagram of the hydroelectric power plant. (Note that the water height of the reservoir is essentially constant.)	233
8.7	Transient response of a hydroelectric power plant.	236
8.8	Piston pushing an incompressible fluid through an orifice.	236
8.9	Piston with a fluid head reservoir.	237
8.10	Graph of the velocity-time characteristic.	239
8.11	Graph of the displacement-time characteristic.	240
8.12	Graph of the acceleration-displacement characteristic.	240
8.13	Tapered-pin approach to the design.	241
8.14	Multihole approach to the design.	241
8.15	Multihole acceleration-displacement relationship.	241
8.16	Blood flow into and out of the kidney.	242

8.17	Autoregulated versus rigid vessel flow.	242
8.18	Symbolic drawing of the vascular bed.	242
8.19	Force balance on the arterial vessel wall.	243
8.20	Flow rate and vessel radius versus perfusion pressure for $P_e = 50$ mm Hg.	244
8.21	Comparison between exact and Galerkin solutions.	245
8.22	Pictorial of (a) static and (b) dynamic torque-angle testing of a finger.	246
8.23	Torque-angle behavior of a finger joint.	246
8.24	Transient response of a thumped finger.	246
8.25	A simplified physical model of the Earth.	247
8.26	Pictorial view of a mechanical muscle. This concept is to be used to power a prosthetics device. The device provides a mechanical advantage for a simple torque motor and produces forces and excursions typical of human arm muscles.	248
8.27	Schematic of a load on a forearm.	249
8.28	Simplified diagram of an aircraft in yaw.	250
8.29	Forcing function for the yaw of an aircraft.	250
8.30	Schematic of an early blowgun.	252
8.31	A simple canoe.	253
8.32	Diagram of a chained system of rigid elements under the influence of gravity and friction.	253
8.33	Econorail dragster.	253
8.34	Expanded view of an automotive clutch assembly.	254
8.35	Energy production system for off-ramps.	255
8.36	Flow diagram of proposed energy retrieval system.	255
8.37	Glass house for use in home gardening.	255
8.38	Simplified model of a helicopter.	256
8.39	Ping-pong ball launcher.	258
8.40	Raising the sail on the *Thompsonia*.	259
8.41	Tank dilution problem.	259
8.42	Launching of an underwater rocket.	260
8.43	The Mercedes Benz T-80 Land Speed Record Vehicle. The T-80 was designed by Dr. Ferdinand Porsche in 1937, completed in late 1939, but never tested or run at speed because of World War II.	260
8.44	An automotive transmission showing the various gear ratios.	261
8.45	Typical automotive water pump.	262
8.46	Simplified representation of falling climbers.	264
8.47	Rod undergoing constrained motion.	264
8.48	Physical arrangement of two masses connected by a massless wire.	264
8.49	Variation of drag coefficient for a sphere with Reynolds number.	265

List of Tables

2.1	Property Data for the Resistor and Its Heat Transfer Coefficient	55
2.2	Physical Symbols and Data for the Source and Body	86
3.1	Linear Approximations	88
3.2	(α, ϕ) vs ϕ, Elapsed Time	107
4.1	Numerical Solution of $dx/dt + x^2 = 1, x(0) = 0$	120
4.2	Application of Euler's Method to the Nonlinear Spring-Mass Problem	123
4.3	Comparison of Exact Solution to Nonlinear Spring-Dashpot with Runge-Kutta Numerical Integration	125
4.4	Property Data for the Resistor and Its Heat Transfer Coefficient	132
5.1	Values of x for Isoclines of Slope α	135
5.2	Initial-Condition Values for the Phase-Plane Solution	146
6.1	Comparison of Perturbation and Exact Solutions	154
6.2	Comparison of Galerkin, Picard, and Secondary Galerkin Methods	171
6.3	Comparison of Galerkin and Exact Solutions for a Bead Oscillating on a Wire	174
6.4	Galerkin vs Finite Difference Solutions to Conduction in an Infinite Wall	179
7.1	Classification of the Four Types of Singularities	193
7.2	Poincaré Index, PI, for the D.C. Generator Example	210
8.1	Manifold Pressure Changes with Engine Speed, Throttle Angle	230
8.2	Property Data for Air and the Fuel Flow System	231
8.3	Relevant Data for the Engine, Air, and Fuel Flow Systems	233
8.4	Physical-Property Data for the System	234
8.5	Approximate Data for the Earth and Sun.	248
8.6	Critical Dimensions for the Thompson Academic Motors, Inc. Model A100 Actuator	249
8.7	Parameter Values for the Aircraft Yaw Problem	250
8.8	Clutch Property Values and Dimensions	254
8.9	LSU Dragster Performance Data	254
8.10	Helicopter Property Data	258
8.11	Numerical Data and Symbol Definition for the Ping-Pong Launcher	258
8.12	Properties of Submerged Launch Rocket	260
8.13	Experimental Data on Porsche/Mercedes T-80 LSR Tire	261

Acknowledgments

One of the greatest influences in my academic career was the late Professor Robert H. Kohr, a faculty member during my graduate studies at Purdue. Much of the material in this text came from his outstanding lectures on nonlinear analysis. I undertook the challenge of writing this book because it was something he would have done had he not tragically perished in an airplane crash at Indianapolis in 1979.

This text would have been impossible without the assistance and contributions of many excellent students and colleagues. Professor Emeritus A.J. McPhate of Louisiana State University (LSU) provided written assistance and significantly influenced my efforts on the material on optimization. Professor Jordan Cox at Brigham Young University reviewed an early manuscript and provided a great deal of material on design methodology. Dr. Warren Waggenspack, Jr., of LSU was a constant source of encouragement and valuable critique during much of this work. Finally, Ric Haag maintained my computing systems at LSU and made my life unbelievably pleasant for many years with his intellect, wit, and computing acumen. He was also the source of much material for the constant-deceleration shock-absorber case study. What would I have done without my friends and colleagues? God bless them all!

I gratefully acknowledge the support shown to me by the Board of Regents of the State of Louisiana and Digital Equipment Corporation. The equipment used to write and test the material on which this manuscript is based was obtained via the Regents grant, LEQSF-86-89, and a Cooperative Endeavor between LSU and Digital Equipment Corporation to support academic instruction in highly interactive modeling and computing.

It is impossible for me to express my gratitude to my family for their patience and understanding when I was writing this in lieu of doing *family sorts of things*. Both of my sons also have done service in reviewing the manuscript and working the problems. Perhaps as this goes to press, I can begin to regain some of this lost time with them.

Even though I have received help from many quarters, I must take full responsibility for the contents of this text, errors and all. Goethe once stated that "A man's errors are what make him amiable." I hope that you, the reader, don't suffer from my generally amiable nature.

CHAPTER 1

The Design Process

The word *design* has very different meanings to different people. The Latin word from which it is derived is *designare*, constructed from the two roots *de*, meaning from, and *signare*, to draw or mark. Thus, the historical meaning of design has definite links to mechanical drawings and artistic renderings. Some of the earliest machines in the form of clocks and gears combined both art and technology; Gerbert invented weight-driven clocks around A.D. 990 [31]. Leonardo da Vinci (1452–1519) was undoubtedly the most influential designer of his period. His artistic contributions are perhaps better remembered today, but he was responsible for many advanced engineering ideas of his time. A major portion of Leonardo's income was from his employment as a military engineer, and his sketches of military assault devices and weapons are certainly impressive. It is an interesting part of this linkage between design and drawing that, in the early 1500s, Leonardo was often granted the dual title of Painter and Engineer by his sponsors. He readily incorporated gears and spring mechanisms into his sketches, but there were no machines for fabricating such systems nor were there materials one could rely on for such machines. What is important, however, is that Leonardo had a conceptual framework for these machine elements that allowed him to incorporate them into his drawings or designs. He recognized the ability of springs to store energy and conceived of machines to use this stored energy. Most of Leonardo's ideas were so advanced that they could not be constructed even in prototype or model form using the technology of the period [53].

It is certain that there is a component of art in design, for both are creative processes. Anyone who has been involved in the design process is aware of the demand on one's creative and imaginative talents. In mechanical design, the connection of drawing and making plans is especially strong, but, unlike the Renaissance days of Leonardo, there is now much more to the design process than making drawings. Leonardo recognized the connections between our perceptions and what we call knowledge. He wrote that "All our knowledge has its origin in our perceptions." One must always be aware of the tenuous and fragile nature of *knowledge*, which sometimes will metamorphose into some new *truths* as our concepts improve and evolve. The concept that all materials exhibit an electrical resistance was shattered by the discovery of superconductive materials. Thus the old knowledge of the behavior of electrical conductance had to be discarded suddenly for a new model that included this new behavior. This concept of superproperties now has invaded our perception of natural phenomena, and science has embraced them readily. The influence of this broadening of our perceptual model of reality subsequently will color all of our studies of natural processes.

1.1 ELEMENTS OF THE DESIGN PROCESS

Engineering often is referred to as the applied arm of science. Knowledge about natural phenomena is used to create new devices, systems, and processes to benefit mankind. The creation of such products involves understanding the needs or motivations behind them, the behavior of the relevant natural phenomena, and an understanding of how to take advantage of this behavior to meet specific needs. This process of creation typically is referred to as *engineering design*. It is the process whereby engineers devise useful products for society.

An integral part of the design process is the development of mathematical models, not only of the natural phenomena involved in the design, but also of the system configuration or the geometry of the products. The essential element of engineering design is thus the ability to synthesize useful designs from these geometric and natural-phenomenon-based models. This ability to synthesize or use the inherent knowledge captured in the mathematical models to create new products is part of the essence of human creativity.

In standard engineering design textbooks, design is broken into distinct phases such as problem definition, ideation, solution specification, design analysis, and evaluation. Although it is useful to consider each of these elements as distinct and apart from the others, design is a continuous process. These phases of design are dissected and discussed in the following sections; the processes to be emphasized in this text are those related to the creation and analysis of mathematical models in design. It is hoped that additional insight into design will evolve in the mind of the reader. The specific subdivisions of the design process used here have been described generally by many others including Koberg and Bagnall [23] and Shigley [39]. Their concepts are included, but they have been updated, compressed, and modified to fit this text. Design has been described as an iterative process as depicted in Fig. 1.1(a). The engineer moves through the flowchart like a computer through its program, always evaluating whether the conditional clauses (functional requirements) are satisfied, and branching and repeating steps where appropriate. Suh [47] characterizes the design process as being more like a mapping, as shown in Fig. 1.1(b). This represents design as a "mapping process from the functional space to the physical space to satisfy the designer-specified functional requirements." However the design process is presented, the common thread seems to be identification, definition, ideation, selection, analysis, implementation, and evaluation. Engineers make their

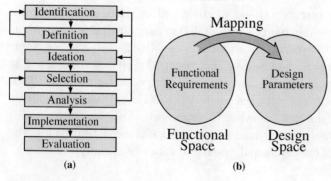

Figure 1.1. Engineering design as (a) an iterative process and (b) a functional mapping.

1.1. Elements of the Design Process

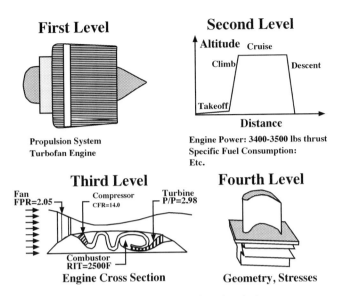

Figure 1.2. Levels of abstraction in engineering design.

decisions on the basis of modeled behavior, needs, and constraints. During the early stages of the design process, an engineer generally will consider many different abstractions of the design problem. Even the definition step is an abstraction, probably one of the most important. An example[1] will best illustrate this concept. Suppose that an aerospace company has identified a gap in the market for a new propulsion system, i.e., an engine. This would be the *first level* of abstraction. The company then would analyze the market potential and specify approximate target values for system parameters such as thrust, fuel consumption, weight, size, and cost. In this manner, a *second level* of abstraction has been defined, one that is more refined than the first level. Figure 1.2 depicts this process of creating levels of abstraction or, conversely, levels of detail. A preliminary design team would now take this second level of abstraction and refine it further by identifying the different parts of the propulsion system and adding the technological knowledge (models and parameters) essential to this new engine. These different parts of the system typically would be the fan, compressor, combustor, turbine, and afterburner. The thermo- and fluid-dynamic models for these elements contain the relevant physical parameters including pressures, temperatures, compression ratios, Mach numbers, and others. These comprise the *third level* of abstraction of the model. Finally, each of the subelements of these systems would be described in terms of their parameters to achieve a deeper level of detail. This results in the definition of shaft diameters, turbine and compressor blades, and combustion chamber sizes and orientations. This represents the *fourth level* of abstraction.

As each level of abstraction is developed, the behavior of the system components at this level of detail may be modeled. At the fourth or fifth level, it may become necessary to include the stress concentrations at the root of the turbine blades or the behavior of molecules in the grain structure of single-crystal casts of these turbine blades. The design

---

[1] Personal correspondence on 06/01/92 with Dr. Jordan J. Cox, Brigham Young University, Provo, Utah.

becomes less and less abstract as one goes deeper into the levels of abstraction and at each level, every detail must be specifically defined.

Engineering design therefore is linked intimately with mathematical modeling. It includes the process of defining the essential levels of abstraction describing the design, but molecular models are not necessary to every design. Thus, the designer must decide on exactly how much detail is necessary and the levels of abstraction required to obtain this information. Only after comparing the model's behavior with that of a physical prototype can the modeling assumptions, which are directly dependent on the level of abstraction used, be verified fully. With experience, the modeler gains added confidence in the models, and subsequent designs may not require full experimental verification unless new and untested frontiers are being forged, e.g., use of new materials, dramatic alterations of the thermodynamic process variables. It is useful at this point to identify and discuss the different stages in a typical design process. This is undertaken in the following sections.

1.1.1 Identification

Any design process must be initiated. This is usually the result of some identifiable problem. However, it may result from the need to correct a previous design flaw, a need to develop a new product with new capabilities to remain competitive, or sometimes a whole new design challenge resulting from an imaginative idea.

Note that many designers fail at this stage, because they have a preconceived notion that proceeding is doomed to failure. At other times it is because they don't see the design problem as "their job." In this phase of the design process, it is essential that one fully accept the challenge as a personal one, and then commit to its resolution.

Shigley [39] indicates that there is a distinct difference between the recognition of a need and the identification of the problem. The problem is more specific. If the need is for cleaner water, the problem might be that of reducing wastewater seepage, industrial dumping, or contamination from natural sources.

1.1.2 Defining the Problem

Definition of a problem is the second step toward its solution. A well-posed design problem is an almost nonexistent entity. For small design problems, it often is the solution. More often, it is recognized that there is a problem, but exactly what the problem may be is not known. There are other times when there is indecision about whether a problem exists or not. The first step, then, is to recognize that a problem exists and to mentally commit to its resolution.

The step of defining the problem also begins with the gathering of information and the attempt to formulate just what the problem really entails. Usually, this process results in an over- or underspecification of the problem. If a problem is overspecified, the constraints on the solution of the problem are so severe that no solution is possible. An underspecified problem is one that is so loosely defined that there are an infinite number of possible design alternatives but no means to determine which to choose. Like the porridge in the fairy tale of Goldilocks and the three bears, the problem definition has to be *just right*. There is, therefore, an evolutionary aspect to the definition of a design problem. As the designer gains experience, this process becomes shortened and more precise because of the expert knowledge brought to bear on the recognition and importance of various subareas. As an example, a designer working on a race car who has experience in solving heat transfer problems in heat exchangers will be able to recognize the relevant heat transfer problems in the vehicle design and remember previous solutions. This is true, even though heat exchangers are but a subset of the overall race-car design problem and his previous experience may not have involved any race-car design.

Suh [47] identifies this phase as "clearly one of the most critical stages in the design process." Poor problem definition often leads to unacceptable or unnecessarily complex solutions. When students are presented with the problem of designing a can crusher, they never consider other solutions to the underlying principle. Here, one must back away from the problem that biases one to specifically *crushing cans*, and look at the more general requirement of *reducing the volume of the can*. By abstracting the problem until one gets wording that is unbiased, but still specific to the problem, one frequently can gain new insights and solutions. If the abstraction is too great, the problem might become too general, and a simple can crusher becomes a "save the world" problem.

1.1.3 Ideation

The training that engineers receive during their academic careers in refining their imaginative or creative abilities is generally insufficient and inadequate. Creativity and innovation in design are, perhaps, areas of design that are the most challenging and the most rewarding. It is a challenge to our ingenuity and our intellect to come up with new and novel insights related to the design that satisfy all of the design constraints. The designer should not be satisfied with a single design alternative, however, but should realize that there are often *many* ways of achieving the major goals of a design. In this phase of design, ideas should be permitted to flow freely and should be recorded for more intensive scrutiny, consideration, and analysis at a later time. Major design decisions are to be avoided at this juncture, and one should merely seek to get new ideas recorded. Although brainstorming sessions are not a common practice, they often stimulate new ideas and begin the real process of taking a fresh, new look. One often can get new ideas by going to the library, reviewing patents in related fields, or even by doodling on a sheet of clean paper. If some method of stimulating your design imagination appears to work for you, you should endeavor to perfect and hone it to a fine art.

1.1.4 Selection

The purpose of a design process is to produce a system, procedure, or device that answers some need. The selection of a proposed solution is a definition or selection of a candidate system or device to answer the specific needs of the design problem.

The selection of a reasonable solution to a design problem involves making choices. This is seldom easy and often must be prefaced by eliminating improper design constraints, defining obvious choices, and deciding on the scope of the problem to be considered. If the problem can be reduced in dimension by any manner, this is the proper stage to do so. If the problem being addressed, for example, involves brakes on an automobile that squeal when they are used, one should not attempt to redesign the entire vehicle. Remember that you are making work for yourself if you do not eliminate all inappropriate design avenues at this juncture because the next step in the design process involves the development of a full model of the design alternative(s) for analysis.

In the selection process, the underlying hypothesis is that there is some mechanism for determining that one design is better than another. What is not stated, however, is that the designer is the one who formulates the criteria for what is considered good or bad.

One of my colleagues once stated: "The best engineering designer is a lazy engineer.[2]" His message was simply that one can avoid work by being clever and finding a better way to do something. He postulated that many of society's major advances have been motivated in this way. Although some do not subscribe wholly to this thesis, there is a germ of truth in such a concept. I have had many personal experiences of looking for a method or mechanism to simplify my work and make it easier and more enjoyable.

1.1.5 Analysis

This is perhaps the most technically difficult aspect of the design process. It involves the description of all of the design alternatives in mathematical terms, thereby allowing the engineer to simulate the operation of the design and to discover any fundamental flaws in the reasoning of his preliminary designs. Much of what is done in this phase of the design process is termed mathematical modeling.

There is much debate over the subject of physically based modeling. Some claim it an impossible art to teach, others attempt to abstract the process using bond graph theory, whereas others attempt to focus on understanding basic concepts and learning to apply them. In this debate, there are no clear winners. However, there is no question that if one has a clear understanding of fundamental engineering knowledge, one has the potential to become a good modeler. Without this knowledge, even the keenest intellect with a natural instinct and talent for modeling will fall short of his or her potential. One can make a similar statement about mathematics. Without a solid grasp of the fundamental constructs and verbs of the language of mathematics, an individual with a natural talent for understanding nature and creating unique and intuitive designs cannot properly express his or her ideas, or investigate them sufficiently to attain their full potential.

In the latter stages of analysis in the design process, through the perspective of mathematics combined with modeling, the most appropriate design will have become obvious and the analysis then will center on the optimization of the design. Many engineers work for most of their professional lives without realizing that their efforts to minimize the cost of a design also should include the cost of the analysis. For many industries, this is the highest cost aspect of the design process, both in manpower and computing.

Another aspect of the design of a product might involve aesthetic factors that, to date, have no mathematical basis. In addition, it is often desirable to involve a full graphic simulation of the product to enable the design to be fully evaluated from all vantages.

There are many levels of abstraction in the mathematical modeling of natural phenomena. The appropriate mathematical model is determined by some agreed-upon level of abstraction. Referring again to the example of a propulsion engine, it would be of little value during the preliminary design stages to do microstructural models of components because the overall engine system will not have been determined fully. The appropriate

---

[2] Professor A.J. McPhate, Department of Mechanical Engineering, Louisiana State University, ca. 1975.

model to be used during these stages would be a thermo- or a fluid-dynamic model that predicts properties at various abstract locations (e.g., pressure, temperature, Mach number at the combustor exit) in the engine. Once the global design parameters have been fixed, one can reduce the level of abstraction and focus on a deeper level of detail.

Often the level of abstraction of the definition of the design can be determined by the description of the geometry. Simple topological maps of the geometry of system configurations indicate a high level of abstraction and, consequently, the mathematical models will be more abstract. Detailed three-dimensional component geometry dictates that the associated mathematical models will be of a more detailed, well-defined type, e.g., finite element or finite difference models.

The development of appropriate mathematical models is of prime importance in the design process. It is from these models that the information essential to the evaluation of the success of a proposed design is derived. This is typically the most expensive phase of the design process because development and solution of detailed mathematical models require many man and computation hours.

1.1.6 Implementation

In this phase of design, the initial prototypes are constructed. Final design decisions are made that might improve the design, make it easier to manufacture, reduce the cost, or simply make it more attractive to the consumer. Through the use of interactive-graphics-based computer simulations, it is becoming more difficult to differentiate this step of the design process from the analysis phase. It is now commonplace for a complete simulation of a system to be used to evaluate a design and to test different scenarios in an attempt to break the design or uncover its weaknesses. The creation of virtual prototypes that can be tested functionally in a virtual application is a powerful paradigm.

1.1.7 Evaluation

Every design must be tested. As our mathematical models improve, it will become practical to test every aspect of a design using computer simulations. Often, when the prototypes are built, new design alternatives are immediately apparent, small detail changes are evident, and obvious errors finally become obvious. Frequently, this phase of design leads immediately back to an earlier phase where redesign becomes necessary before releasing the design for production. Hardening all design parameters is the real product of this aspect of design. It is the reward of all of the previous efforts.

Usually, this phase of the design sequence culminates with the actual construction of a few prototypes and validation of the design through direct testing and comparison with the computer simulations. As we gain more knowledge and experience with physically based modeling, the necessity for complex and expensive benchmarks will abate, and the virtual world of simulations will become accepted practice.

1.2 THE COMPUTER AS A DESIGN TOOL

There are many uses of the computer as a design tool, and it is practically impossible to list them all. It is essential, however, to recognize that (1) the computer is a tool and, even with the most elegant software, is not a panacea for all design problems; and (2) the computer may be used in almost every phase and aspect of design. Some valuable ideas include

Office automation. The engineer spends 90% of his time doing things other than design. A great fraction of this time is not even spent doing technical work. During this time, the designer must manage his calendar and his telephone calls; write reports, proposals, letters, and other documents; communicate to his peers at the technical and at other levels; organize his ideas, and perform many other tasks for which computers are now becoming recognized as valuable tools.

Technical assistance. The computer generally offers many utilities for assisting the engineer in his work. These range from spreadsheet modeling tools to algebraic manipulators and programs to aid in the definition of three-dimensional geometry.

Networking. The use of network connectivity for professional workstations allows the engineer to communicate readily with technicians or other subordinates, his supervisor or other managers, and other engineers and professionals around the country and the world. This assumes that his company has seen fit to avail itself internally of electronic networking as well as external connections to the growing networks in the United States and other countries. Many experts are available through the networks for consultation and there is a growing amount of database information available on these networks.

Graphics. The ability to organize large amounts of information in the form of graphical presentations is enhanced greatly by the workstations of today. This may be in the form of a simple plot of one variable versus another or it may be a multidimensional plot in which colors and other visual clues are used to relate information to the viewer. Other uses include interactive three-dimensional visualization of a design, complete with surface finishes, colors, and other material properties to create realistic images. The real value in this form of presentation is that the human brain has developed a robust system of pattern recognition circuits that are designed specifically to process visual images.

1.3 SUMMARY

Highly abstract definitions of designs and their accompanying mathematical models usually can be posed, developed, and solved by a single designer. As one proceeds to deeper level of abstraction, however, more detail and expert knowledge are needed. Design specification becomes more intense and mathematical models become more complex. This often requires that whole departments in a company be devoted to the development of a component of some larger system at a higher level of abstraction. This may include the definition of geometry and the resultant mathematical models for heat transfer, stress and strain, vibration, or a myriad of other types of analyses. Complexity requires specialization, but it is important to maintain an understanding of the evolution of the abstractions of the proposed design because there is a great deal of inherent information contained in this process.

Design decisions frequently move the specification to deeper levels of detail. In this process, the specifications become less and less abstract and more and more specific. The results of a mathematical analysis at a deep level of detail may indicate that a design

1.3. Summary

decision based on an assumption made at a more abstract level was incorrect. Maintaining a record and understanding the reasons for moving to deeper levels of detail provide mechanisms for retracing previous steps and refining the abstractions. Invalid assumptions can be identified and corrected, occasionally producing a new structure of detail.

The keys to successful design still are not understood completely. Creativity, attention to detail, good data management, mathematical prowess, and other attributes all seem to play a part in this process. An essential ingredient to improving one's approach to design is an understanding of how to abstract design specifications and create mathematical models. Determining the proper type of mathematical model can make an enormous difference in whether a designer can identify the critical aspects of the design. Engineers too frequently jump to a level of great detail and attempt a large-scale finite element analysis when a simpler and more abstract model might have allowed refinement of the geometry and reduced the computer analysis. Not only are money and time wasted, but more appropriate alternative designs may have been missed because of the brute-force approach.

CHAPTER 2

Mathematical Modeling

All humans model. We build these models within ourselves as a template for what to anticipate under the varied conditions of our daily existence. In attempting to harness some of this native skill for the purposes of engineering design, however, we must begin to recognize what models are, how we build them, what their limitations are, and how we can improve them to best suit our purposes. In this text, the words *design* and *modeling* are used frequently. Modeling in the context of this book is hereafter taken to be the use of a *mathematical* representation or formalization of some natural behavior or phenomenon. This may be the behavior of a complex system (e.g., the heating and cooling of the earth) or something simplistic (e.g., the thermal response of a thermocouple). The term modeling has its roots in the Latin word *modulus*, meaning "small measure." Thus, a mathematical model is, in some sense, a measure of the behavior of what is under study. Also implied in this word is that the model is smaller than the original, and that there is a similitude in its actions.

Design is very different. It is the process of synthesizing some item of purpose, and then deciding what values should be assigned to the quantities that define the item, be this a constant in an approximating polynomial, the width of a piston ring in a diesel engine, or the hourly itinerary for a rock-band tour. Generally, modeling and analysis are used in the definition of a design.

To questions about their definition of design, many engineers respond with a concept that design is a process leading from an idea to a tangible object or process. Others feel that it is merely the application of one's intellect to the solution of a definite problem using engineering principles. All agree, however, that the introduction of computing engines and computer graphics has had an astounding positive impact on design. At one point in the history of engineering, computing systems were leading engineers away from the traditional drawing-based practice of design. With the advent of interactive graphics, this trend has been reversed, and full-color realistic renderings of design ideas now can be viewed from any perspective. Stress levels, pressure fields, temperatures, and other information can be readily mapped onto these renderings to aid in the process of design evaluation. In addition, as the power and sophistication of computing systems increase, it becomes feasible to implement complex mathematical representations or math models of the systems under analysis. This places an increased burden on the mathematical tools of the engineer and, in part, is the reason this text has been written.

The use of mathematics to model the real world is the basis for modern engineering. It is now standard engineering practice to formalize the design of some system and then simulate the loads, stresses, flow patterns, thermal gradients, and other physical phenomena. This is done so that the engineer may evaluate the efficacy of the design, or optimize

Chapter 2. Mathematical Modeling

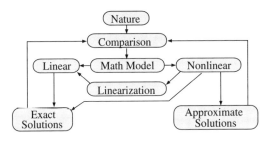

Figure 2.1. Schematic representation of mathematical modeling of real problems.

it relative to some criteria defining what is considered by the engineer to be a good design. This procedure, depicted in Fig. 2.1, shows two distinctly different approaches taken when modeling systems mathematically, and are broadly categorized as either linear or nonlinear systems. For linear systems, exact solutions to the model equations are possible, and many methods may be used without trepidation. The defining model equations for nonlinear systems, however, are not as amenable to solution, and this text attempts to present many methods for their solution. Further, note that modeling is not an open-loop process. After one has modeled a system and generated a solution, it is essential that a comparison be made between this solution and the natural phenomenon that is under study. If the two are unacceptably different, one must improve or reformulate the model and try again.

Consider the simple task of modeling the uniaxial stress-strain relationship of carbon steel. This relationship is depicted graphically in Fig. 2.2, which shows the resulting strain for a given applied one-dimensional stress. The region near the origin is certainly where most engineering designers attempt to constrain their designs' operation, for here, stress S and strain ϵ are linearly related and the material is not in danger of immediate failure. The mathematical model for the behavior of the material in this region therefore is written as

$$S = E\epsilon \tag{2.1}$$

where the constant of proportionality, E, is the modulus of elasticity (Young's Modulus) of the material. This simple relationship is valid only over the region O-A. At higher stress levels, a different modeling approach must be used to model the nonlinear behavior of the material. A simple Nth-order polynomial might be used, such as:

$$S = E_1\epsilon + E_2\epsilon^2 + E_3\epsilon^3 + \cdots + E_N\epsilon^N. \tag{2.2}$$

Unfortunately, even this modification can be used only to extend the coverage of the math model to the region B-A. Above this, permanent deformation occurs and a residual strain is left in the material even after the load is removed. Once this has occurred, strain becomes dependent on the complete history of the loading of the material. Representing

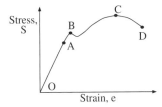

Figure 2.2. Stress-strain behavior of plain carbon steel.

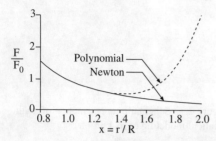

Figure 2.3. Comparison of Newton formulation (solid line) and polynomial (dashed line) representations.

physical phenomena using polynomials generally does not represent a viable approach to mathematical modeling; whenever possible, equations that represent the behavior should include real, meaningful parameters. It would have been far less important had Newton[1] described the gravitational attraction of two bodies separated by a normalized distance

$$x = \left[\frac{r}{R} - 1\right]$$

using a polynomial such as

$$F = \frac{F_0}{R^2}(1 - 2x + 3x^2 - 4x^3 + 5x^4) = \sum_{i=0}^{N} A_i x^i. \tag{2.3}$$

In this representation, the coefficients A_i have no physical meaning other than that they are the first $N = 5$ coefficients of the Taylor-series representation of $1/r^2$ about $r = R$, or $x = 0$. The value of a real physically based mathematical model is obvious when one examines Newton's law as shown in Eq. (2.4):

$$F = \frac{\gamma M_1 M_2}{R^2} \tag{2.4}$$

Here, unlike the polynomial representation of Eq. (2.3), the importance of all of the physical parameters is immediately obvious and accessible. When one graphically compares the two formulations [Eqs. (2.3) and (2.4)], it also may be that polynomial formulations fail outside of a given range, as shown in Fig. 2.3.

It is a fact that, in general, the simulation equations for most real physical systems are nonlinear; the real world is nonlinear. As depicted in Fig. 2.1, there are three basic procedures for finding solutions to these equations: (1) Obtain an exact solution; (2) obtain an exact solution to an approximate (crude) problem; and (3) obtain an approximate solution to a refined, exact problem statement. The determination of which is the best method lies in a comparison of the two different solutions to the behavior of the real-world situation being modeled. There are no easy answers to this question or universal approaches to the solution of such nonlinear simulation equations. It is also interesting to note another property of nonlinear solutions; they often exhibit particular solutions that are impossible to obtain from similar linear systems. They even occasionally behave in manners not even suggested by their linear counterparts. This presents both a problem for the modeler attempting to understand and gain confidence in his or her solutions as well as an extremely interesting and challenging component of nonlinear analysis.

---

[1] Sir Isaac Newton (1642–1727).

2.1 SPECIAL BEHAVIOR OF NONLINEAR SYSTEMS

To aid the modeler in better understanding and interpreting the results of analysis of nonlinear systems, some of the features of these systems are enumerated in this section. This is not an exhaustive or exclusive list, but it represents many of the features common to nonlinear system behavior.

1. The principle of superposition does not apply to nonlinear systems. If an input x_1 results in an output y_1, and y_2 results from an input x_2, then, as depicted in Fig. 2.4,
   (a) for linear systems, when the input is $x_1 + x_2$, the output is $y_1 + y_2$;
   (b) for nonlinear systems, this is not the case.
2. Frequency components appear in the output, even though they are not present in the input (see Fig. 2.5). These frequencies are often some integer fraction or multiple of the forcing function or the natural frequency of the system. This is not always true, however, and sometimes the frequency content of the output appears arbitrary. At other times, the system may appear to behave in a totally chaotic manner.
3. The stability of linear systems is determined only by the differential equations describing the system, not the initial conditions, boundary conditions, forcing functions, etc. The stability of nonlinear systems is determined by all of these.
4. The steady-state output of a nonlinear system may be an oscillation of fixed frequency and amplitude, often attained regardless of the initial state of the system. These steady-state oscillations are termed *limit cycles*.

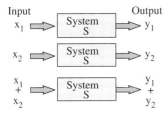

Figure 2.4. Response of a linear system (principle of superposition).

These and other facets of the behavior of nonlinear systems are discussed in subsequent chapters along with methods of solution of the nonlinear mathematics representing these systems. Be certain to look for these important properties of nonlinear systems in the examples given and in your solutions to the problems associated with each section. In Chapters 3 through 8, many examples of such system models are presented. The governing equations for these models are solved using various methods, but little mention is made of how these models were created or the inherent assumptions within them. In subsequent sections of this chapter, the basic constructs of modeling are discussed, and several models are developed and presented. Solutions are shown that can be compared to those obtained using other methods. The reader is strongly encouraged to attempt other solutions using the techniques presented in subsequent chapters.

Figure 2.5. Comparison of input and output for linear and nonlinear systems.

2.2 MODELING BASICS

It is very difficult to teach someone how to mathematically model physical systems. The most common practice is for college professors to teach modeling indirectly through the many examples, derivations, and other classroom materials that portray models without any emphasis on the modeling methods themselves. The actual practice of modeling seldom is taught. This is especially puzzling in light of the depth to which mathematical modeling is embedded into engineering and science. Further, psychologists agree that people are continually forming their own personal mental models of nature from cradle to grave. The key step in creating a mathematical model is expressing these mental models formed by ideas and concepts into useful mathematical expressions. From this vantage, mathematics is seen as a formal language that allows expression of abstract ideas in a concise and complete manner. The prerequisites for applying this transformation are a broad knowledge of elementary physics and engineering and a good grasp of mathematics and calculus. With these tools and the experience born of practice, anyone can build useful mathematical models.

In creating a mathematical model, there are several important issues to be considered. These include the definition of

1. the model's purpose or objective, which includes a statement of who will use the model;
2. available resources (time, money, computing, software, etc.), which also includes limitations or constraints placed on the model;
3. basic knowledge relevant to the model, data on properties, or other behavior.

These statements often will guide one into specific kinds of models or to models having different levels of complexity or flexibility. In building a mathematical model, it is noteworthy that one is also *designing* the model. Thus, creating a math model for the purposes of a design and analysis is, of its own merit, a design. The term analysis continues to appear, frequently linked to design and to solutions of model equations. The word is derived from the Greek *analyein*, meaning to separate or break into constituent parts. Analysis, therefore, is defined here as an examination of the elements of a system and their relationships.

In Sections 2.2.1 through 2.2.5, the various aspects of math modeling are described. One classical approach in modeling involves the technique of applying the *razor of Occam*. This technique was developed in the fourteenth century by an English philosopher, William of Occam, and involves how one perceives reality, i.e., how one builds a mental concept of reality. The Germans coined the word *gedanken*, or *mental experiment*, for this way of intellectually testing and sifting material and facts. In this method, reality is virtually squeezed through the hole of human observation. The size of the hole is symbolic of the level of detail to be included in the model. Each piece of knowledge thought to be of importance to the model is tested to determine its relevance. If it is found to be of vital importance, it is then spared the razor of Occam. If it is unimportant, it is stripped away by Occam's razor. This method has been used by math modelers for many years, including Heisenburg,[2]

---

[2] Werner Heisenburg (1901–1976).

2.2. Modeling Basics

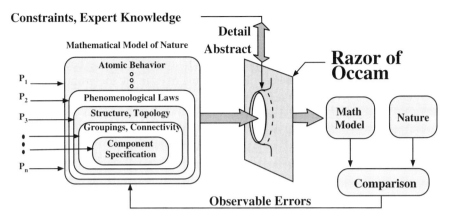

Figure 2.6. Application of Occam's razor to modeling.

Schrödinger,[3] and Dirac,[4] who used the method to demonstrate that position and velocity could be stripped from the model of elementary particles in lieu of a quantum description of the states of the particles. The value of this technique occurs during the initial definition of the model. Any aspect of reality that doesn't contribute to the purpose of the model is omitted. In addition, as constraints are added to the model, the razor of Occam must, of necessity, become more active, stripping away nonessential detail. This concept is shown in Fig. 2.6. Note that the greater levels of abstraction survive this process better than the greater levels of detail. It also is inferred that the parameters associated with each level of abstraction, P_i, also are stripped away by the razor of Occam.

Through the use of mathematics, one can invoke the power of abstraction and the power of analytic tools to probe the behavior of math models. Here, the power of abstraction derives from the syntax of mathematics: rules of symbolic manipulation of variables that permit operations such as addition, subtraction, differentiation, and integration. Digital computers have programs that permit symbolic expressions to be manipulated using the language of mathematics, greatly speeding the associated operations and ensuring accuracy. In math models, these symbolic elements include real physical model parameters and properties. Thus, one always should defer the use of numeric values for these entities until the later stages of the analysis to take full advantage of abstraction.

On a historical note, the first mathematician to attempt to use symbolic algebra is believed to have been Diophantus (ca. A.D. 275) in his treatise entitled *Arithmetica* [43]. This treatise is only known through references in more recent manuscripts dating from the thirteenth century. Diophantus wrote his equations much as we do today, but his symbols would not be familiar to the modern modeler. It is also noteworthy that the use of arabic numerals (0 through 9) as symbols for the size of something were not in widespread use in Europe until around 1500 [46]. This was also the period when modern abstract symbology for equations evolved. Thus, one reasonably could claim that modern math modeling, in which knowledge is stated in abstract symbolic form using a mathematical language, has only been in existence for approximately 500 years.

---

[3] Erwin Schrödinger (1887–1961).
[4] Paul A.M. Dirac (1902–1984).

2.2.1 Purpose

The definition of the purpose or objective of a math model is essential to the success of the model. If one were to model an automobile by adding up the dollar cost of every component, the model would serve an economist seeking to estimate the profit margin of this product quite well. This same model would not be fitting for an engineer seeking to estimate the mean time to failure based on the stresses within the various elements of the vehicle. Don't forget to invoke Occam's razor as early in the modeling process as possible, even when formulating the purpose of the model. When in doubt, don't try and make the most sophisticated model of a problem first. It is usually more valuable to start with a simple model having a limited purpose, and then to improve upon it as needed.

This is perhaps the most difficult aspect of modeling, for it has been stated by others that most problems are solved once they have been properly defined. There is a great deal of truth in this view, and one should spend whatever time and energy necessary to properly understand the problem before committing an attempt at modeling and analytically solving the problem.

2.2.2 Resources

Before one can design a mathematical model, one must know what resources are available. This is much like trying to set up a project to accomplish a given goal without knowing what the budget is. The resource list, however, generally should not be restricted to money alone. It should encompass the available manpower and time, the availability of computing resources, software programs, and any other resources or limitations and restrictions on resources. This can have significant impact on the type and detail of the model produced. If one has 15 minutes to create a math model, for example, it will not have any subtle or sophisticated features, nor will it involve the writing of a computer program or doing a detailed literature search.

The general principle of starting simple in model development allows one to expend few resources in developing a first model attempt. This step also helps to assure one that the real problem is being solved, while further refining the modeler's understanding of the problem and including more detail on the elemental natural phenomena involved. Having completed a simple model, new directions for the continuing evolution of the model are often obvious.

2.2.3 Expert Knowledge

An understanding of the basic knowledge essential to the model includes expert knowledge. At times, this may be a simple tabulation of facts such as the list of available bolt sizes and their allowable stresses, weights, and dimensions. Or, it could be the load-carrying capacity for an automobile tire at different inflation pressures, widths, and diameters. In many such cases, the variables are not only discretized, they are restricted to the list of available diameters, widths, etc. This often discretizes the design alternatives as well. In other situations, the expert knowledge might represent experimental data such as the effect of Reynold's number on the friction factor for pipes of differing roughness as summarized in the Moody diagram [29]. This also points out that many models that otherwise could be thought of

2.2. Modeling Basics

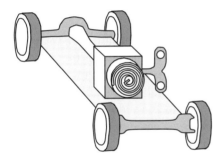

Figure 2.7. Spring-driven toy car.

as continuous in nature become discretized by real-world considerations. The number of bolts on a flange must be an integer, but the trade-off between this number and a selection of bolt diameter from standard bolt sizes dramatically reduces the choices allowable.

Part of the expert knowledge might be an understanding of the solution space. If, for example, the knowledge that the density of a working fluid in a process must be real, positive, and no greater than 20 g/cm^3 could be valuable information to a designer writing a computer program to optimize the process by varying the working fluid's density. Such a program could, unless constrained, find an optimum using a density value that is imaginary or negative. Neither of these is found in nature. It is only through the expert knowledge of the programmer that such solutions are discarded by the search algorithm.

One of the important aspects of designing a mathematical model is deciding which parameters are important to the model. In some instances, there may be parameters that are not independent and their relationships are secondary to the central model. As an example, if the spring-driven toy car of Fig. 2.7 were being modeled, the diameter, height, and thickness of the wound coil spring all would be assumed to be important variables that influence the spring's ability to store energy. These physical parameters could be lumped together to form a function that predicts torque for any given spring shaft angle. Whether this functional were linear (e.g., a single spring constant) or nonlinear with respect to the shaft angle would have to be considered next. It is obvious that the number of free parameters is reduced by this simplification and that the chances for algebraic errors are reduced. If, however, the designer is attempting to constrain the design with respect to the weight of the toy, this simplification eliminates the very geometric parameters that allow computation of spring weight.

The initial formulation of the equations that govern a model often is done in verbal statements. This applies to the statement of the governing principles as well as to the constraints of the situation. The formulation may be based on an energy analysis, a cost analysis, or any other governing principles appropriate to the system being modeled. Each term in these verbal statements then must be converted into mathematical formulas. You must constantly and carefully check at each step to ascertain that every term in your equations has the correct units, and that the algebraic sign of the term makes sense in the overall context of the equations. At the end of this process, the model is therefore a compendium of what is known about the behavior of the modeled system. It may not contain all that is known about the physical system, but it should contain all relevant terms. The model construction is therefore a statement of expert knowledge. It is your task to discern that all of the statements are true. Mathematics is a language just like English, French, or any other. It cannot prevent you from making false or misleading statements.

2.2.4 Model Validation

Every mathematical model must be verified before it can be of real value in design analysis. This is sometimes done using prototypes and experimental verification of the model's behavior. In some instances, it is possible to consider special limiting cases to ensure that the model has the proper behavior. In others, the mathematical models are so complex that true validation is almost impossible. If, however, the model were started as a simple model and then refined, the underlying basic model elements could be validated early in the modeling process and only the improvements then could be termed questionable.

One piece of expert knowledge, for example, that applies to the mathematical modeling process itself is that every term in a mathematical expression must have the same units as every other term. This consistency check is done immediately after deriving a model of a system to ensure the validity and soundness of the model representation.

2.2.5 Constraints

The definition of the model's constraints is an essential component of the model. This definition should be formalized in mathematical terms; one often learns a great deal about constraints just in attempting to write them down. Again, Occam's test infers that adding constraints to the model renders it less flexible and/or powerful.

2.2.6 Examples of Modeling

In the following, simple models are used to demonstrate the principles defined earlier.

EXAMPLE 2.1: EGGS-IN-A-BOX

As an introduction to the importance of modeling and the issues just mentioned, consider the following (Note: I have big feet):

How many eggs can you get into a shoebox that is 6 in. $\times$ 8 in. $\times$ 15 in.?

In the next 60 seconds, outline a model for answering this question. Do not read further until you have attempted this.

First, 60 seconds is a very short time. This is a limitation on available resources and the model you create will be directly influenced by this. If your thought processes are similar to that of others, you attempted to approximate the size of a generic egg as 1 cubic inch and divide this into the volume of the shoe box to obtain 720 eggs. Since this was a crude approximation, one also might wonder what the effect of a real egg's geometry would be. Second, because of your time limitation, you probably also went straight for a numerical calculation. This shortcut eliminated one of the most powerful features of a model: symbolic abstraction. It is often valuable to convert your conceptual model from one that is strictly numerical to one that includes symbols for the pertinent dimensions or other variables. This is done in the following paragraphs along with a sophistication of the original model.

As a second assignment, attempt to write down in verbal statements on how you would compute the number of eggs in a box. Then convert this to a mathematical formula. Does this method of abstraction provide you with new ideas on how to improve your original

2.2. Modeling Basics

model? Discuss it with your colleagues if you wish, but do not take longer than 5 minutes to complete the assignment. Again, do not read further until you have completed this.

Did you consider the purpose of the model? Did you decide on the most essential elements while stripping all nonessential components? Were you able to use analytic variables in lieu of numeric ones? Did you then check every term for dimensional equality? In all probability, the answer to some of these questions was "no." Generally, the first and simplest alteration most individuals make to their egg model is to make it aspherical. This is done by assuming one dimension slightly longer than the other. A single egg's cylindrical equivalent volume is then $v = ed^2$, where d is the minor diameter of the egg and e is the length of its long axis. Mathematically, the volume of the shoe box may be expressed in terms of its length L, width W, and height H as

$$V = LWH. \tag{2.5}$$

One now can estimate how many eggs would fit into the box as V/v. Assuming that the eggs then can be stacked into the volume evenly, with no fractional eggs on any boundary, the number of eggs then is expressed in words as

Number of eggs = Volume of box divided by the volume of a single egg

or in mathematical terms as

$$N = \left(\frac{V}{v}\right) = \left(\frac{LWH}{ed^2}\right). \tag{2.6}$$

This formula is misleading, however. It implies a continuous relationship between the dimensions of the box and the number of items in the box. Since there cannot be a fractional number of eggs, this can be noted by the equation

$$N = \text{floor}\left(\frac{LWH}{ed^2}\right). \tag{2.7}$$

Here, the floor() function returns the integer of the fraction rounded down. The assumptions used to this point, however, make this type of model a very conservative estimate of the number of small eggs in a large box. The expression given by Eq. (2.7) can be considered as an estimate of the lower bound on the number of eggs. Given more time (resource) and perhaps a digital computer (resource), one could come up with a better, more accurate model.

As noted in Fig. 2.8, the original stacking model of the eggs can be measurably improved upon. Thus, a better model is given as

$$v = ed\left[d\cos\left(\frac{\pi}{6}\right)\right] = \frac{\sqrt{3}ed^2}{2}. \tag{2.8}$$

Figure 2.8. Improved model for stacking eggs.

This increases the original estimate of the number of eggs by 33%. Thus, as more time and effort are expended, one obtains an even greater precision with these math models.

The modeling of any system, whether a simple egg-stacking problem or a complicated technical application, is thus very dependent on the issues raised earlier. As one applies more resources, the sophistication and accuracy of the model improve accordingly. As noted in Eq. (2.7), what was important to the problem also can be abstracted and perhaps used to improve upon the model. Suppose that the question posed before was not "How many eggs can one put into a given box?" but "How can we optimize a box design to maximize the number of eggs it can hold?" A study of Eq. (2.7) shows that the dimensions of the box are not as important as their ratio to the dimensions of the egg. This is, again, a result of the fact that the number of items in the box is not a continuous function of the box dimensions. Thus, one may write

$$N = \text{floor}\left(\frac{L}{e}\frac{W}{d}\frac{H}{d}\right) = \text{floor}\left(\frac{L}{d}\frac{W}{e}\frac{H}{d}\right) = \text{floor}\left(\frac{L}{d}\frac{W}{d}\frac{H}{e}\right). \tag{2.9}$$

The order of arrangement of the ratios may be of consequence, for it is not the integer of the entire expression that is important but, rather, the integer of each term. Also, for given dimensions of box and egg, if one defines

$$I = \text{floor}(L/e)\text{floor}(W/d)\text{floor}(H/d), \tag{2.10}$$

$$J = \text{floor}(L/d)\text{floor}(W/e)\text{floor}(H/d), \tag{2.11}$$

$$K = \text{floor}(L/d)\text{floor}(W/d)\text{floor}(H/e), \tag{2.12}$$

then it follows that the maximum number of eggs is thus determined by $N = \max(I, J, K)$. To maximize this, it is obvious that if one keeps the ratios in these equations to exact integers, the number of eggs is maximized for that volume and stacking method. A secondary effect is that of closer stacking in the third dimension. By alternating the stacking direction from layer to layer, it is possible to further increase the number of eggs in any given volume. To continue this effort, one may add more features to the model, such as (1) a search for the best direction along which to orient each layer of the long axis of the eggs for given values of L, W, and H; and (2) statistical estimates of the minor and major axis lengths of eggs.

This example is very superficial. Hopefully, you have already uncovered some major problems with the model such as the fact that not all eggs are the same size or that there is a practical limit to the number of eggs one can stack before the lowest egg is cracked by the weight of those above it. Even this number will be reduced if the box is to be subjected to anything but the most gentle handling.

EXAMPLE 2.2: MODEL OF A WHALE

Consider the question of how fast whales can swim. This problem seems to be related to energy considerations and fluid mechanics. The question posed involves a prediction of the terminal velocity of a whale at a maximal sustained effort. It is often helpful to sketch a simple figure to assist in forming the concepts and ideas involved. Most engineers are taught to draw free-body diagrams in statics and dynamics, so this is not a new approach. Don't be limited, however, in the information you place in such drawings;

2.2. Modeling Basics

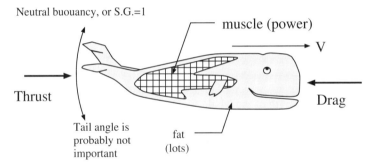

Figure 2.9. Sketch of a whale and some related information.

include pictorials of anything you consider of interest. Figure 2.9 demonstrates such a sketch. The expert knowledge required is in three areas: (1) muscle energy capacity, (2) whale physiology, and (3) drag on submerged objects. The verbal statements of principles in these areas are

1. **Muscle energy.** The power, P, available from a skeletal muscle often has been related to its total mass, m. In words, this statement is

   Muscle mass is proportional to the maximum power potential.

   This can be converted to the mathematical expression,

   $$P = km. \tag{2.13}$$

2. **Fluid mechanics.** A power analysis of the whale can be expressed verbally as

   The power, P, required to move the whale through the water at a velocity V is the maximal thrust, T, times the velocity.

   This is written mathematically as

   $$P = TV. \tag{2.14}$$

   At maximal thrust, however, the whale will have reached a speed at which the drag equals the thrust. In fluid mechanics, the drag often is modeled as

   Drag (thrust) is proportional to the dynamic pressure and the frontal area, A_F. The constant of proportionality is termed the drag coefficient, C_D.

   In equation form, this becomes

   $$T = C_D A_F \left(\tfrac{1}{2}\rho V^2\right). \tag{2.15}$$

   Equation (2.15) is, by itself, a mathematical model of drag. Thus, our models frequently also contain models with a concomitant increase in the detail level of the mathematical description.

3. **Whale physiology.** The amount of muscle mass, m, in a whale's body can be modeled simply. This doesn't really require a great deal of expertise and can be verbally written as

   Muscle mass is a simple fraction of the whale's body weight.

Introducing the additional knowledge that the specific gravity of a whale is very near one, the density, ρ, of a whale is approximately that of water. Therefore, in mathematical terms, this fraction, r, is written

$$r = \frac{m}{M} = \frac{m}{\rho A_F L}, \qquad (2.16)$$

where L is some characteristic length of the whale.

Combining all of the above equations and solving for the velocity yields

$$V^3 = \frac{2krL}{C_D}. \qquad (2.17)$$

Numerically, given the following data,

mass potential, $k = 20.0$ ft-lbf/lbm-s,

mass fraction, $r = 0.25$,

drag coefficient, $C_D = 0.35$,

characteristic length, $L = 30$ ft,

it is possible to compute that the maximum speed of a whale is approximately 30 ft/s or 20 mph. There is very little quantification of whale drag coefficients, average lengths and speeds, and percentage of body weight invested in tail muscle. There are qualitative descriptions, however, and the Right Whale has a very large percentage of its body weight in blubber (small r-value), a shorter, blunter body (high C_d), averages around 50 feet in length and has a top speed around 15 mph. The Fin Whale, by contrast, is the longest, sleekest, and fastest of the whales. It grows to approximately 80 feet in length and can attain top speeds of 30 mph. For these two whales, our theory over-predicts the top speeds at 27 mph and 42 mph, respectively. In view of the gross assumptions made, these are acceptable first order estimates. Although this may be the maximal velocity, one might surmise that this speed is too great for a whale to sustain for very long. It is left as an exercise for the reader to extend this model to estimate the length of time over which a whale can continue at this speed.

The next three sections include some of the tools that are useful in describing physical systems. The method of Lagrange equations is presented first, followed by a discussion of vector mathematics, including the higher-order indicial notation, and concluding with a basic integral conservation formulation.

2.2.7 Method of Lagrange's Equations

The following is a procedure for deriving the governing equations for nonlinear systems having several degrees of freedom. The method generally is used for structural dynamics problems, heat flow, fluid flow, and other energy-related models. It requires defining measures of the potential energy of the system, U, and the kinetic energy of the system, T,

2.2. Modeling Basics

from which a Lagrangian function, $L = T - U$, is derived. The classic Lagrange[5] equation then is written as

$$\frac{\partial}{\partial t}\left(\frac{\partial L}{\partial \dot{q}_i}\right) - \frac{\partial L}{\partial q_i} = Q_i. \tag{2.18}$$

The term Q_i is the virtual work done on the system by all external forces, both conservative and nonconservative. The procedure involves an approximate characterization of the solution applied to the Lagrange equation description of the system. The procedure for this method involves six steps:

1. Model the energy, work, and displacements (strain) of the system.
2. Select an approximation function that captures the character of the solution and satisfies as many boundary conditions as possible.
3. Substitute these functions into the formulas of step 1.
4. Insert the resulting energy, work, and displacement functions into Lagrange's equations.
5. Perform the necessary differentiations and integrations to obtain the governing differential equations.
6. Solve the resulting (nonlinear) differential equations.

In this method, any boundary conditions that are not used in creating the governing differential equation can be used as additional constraints to the final solution. In the Lagrange method, the concept of generalized coordinates, q_i, frequently is employed to describe the system's motion. These coordinates are not necessarily the spatial variables characterizing physical measurements, but are any convenient parameters characterizing the motion. The number of generalized coordinates must match the number of degrees of freedom of the system.

EXAMPLE 2.3: BEAM CLAMPED AT BOTH ENDS

Consider a uniform beam of length L and elastic modulus E that is clamped at both ends. The beam is assumed to have a section moment of inertia I, a cross-sectional area A, and a mass per unit length m. The lateral displacement for the beam is $v(x, t)$ and its axial displacement is $u(x, t)$. The strain energy is thus

$$U = \frac{EA}{2}\int_{-L/2}^{L/2} \left[\frac{\partial u}{\partial x} + \frac{1}{2}\left(\frac{\partial v}{\partial x}\right)^2\right]^2 dx + \frac{EI}{2}\int_{-L/2}^{L/2}\left(\frac{\partial^2 v}{\partial x^2}\right)^2 dx \tag{2.19}$$

and the kinetic energy is given by

$$T = \frac{m}{2}\int_{-L/2}^{L/2}\left[\left(\frac{\partial v}{\partial t}\right)^2 + \left(\frac{\partial u}{\partial t}\right)^2\right] dx. \tag{2.20}$$

The next step involves representing the unknown functional behavior for the system displacements. In this example, the x-dependent functions are termed the *mode shapes*

---

[5] Joseph-Louis Lagrange (1736–1813).

and they imitate the assumed solution behavior, having zero displacements and derivatives at the ends. The lateral displacement, v, imitates the first deflection mode shape, and the axial mode shape attempts to relieve the stresses at points where the slope is large:

$$u(x,t) = \xi_u(t) \sin \frac{4\pi x}{L}, \tag{2.21}$$

$$v(x,t) = \frac{\xi_v(t)}{2} \left[1 + \cos \frac{2\pi x}{L}\right]. \tag{2.22}$$

The unknown time-dependent coefficients are termed *modal amplitudes*. After substitution of Eqs. (2.19), (2.20) into Eqs. (2.21), (2.22), the resulting energy formulations become

$$U = \frac{EA}{2} \int_{-L/2}^{L/2} \left[\frac{4\pi \xi_u}{L} \cos \frac{4\pi x}{L} + \frac{\pi^2 \xi_v^2}{2L^2} \sin^2 \frac{2\pi x}{L}\right]^2 dx$$

$$+ \frac{EI}{2} \int_{-L/2}^{L/2} \left[-\frac{2\pi^2}{L^2} \xi_v \cos \frac{2\pi x}{L}\right]^2 dx \tag{2.23}$$

$$T = \frac{m}{2} \int_{-L/2}^{L/2} \left[\dot{\xi}_v^2 \left(\frac{1}{2} + \frac{1}{2} \cos \frac{2\pi x}{L}\right)^2 + \dot{\xi}_u^2 \left(\sin \frac{4\pi x}{L}\right)^2\right] dx \tag{2.24}$$

The Lagrange equation [Eq. (2.18)] is rewritten in terms of the problem variables as

$$\frac{\partial}{\partial t}\left(\frac{\partial T}{\partial \dot{\xi}_i}\right) + \frac{\partial U}{\partial \xi_i} = Q_i, \quad i = v, u. \tag{2.25}$$

The virtual work function is Q_i, and after the substitutions, this yields the Lagrange equations for this system:

$$m\ddot{\xi}_v \int_{-L/2}^{L/2} \left[\frac{1}{2} + \frac{1}{2}\cos\frac{2\pi x}{L}\right]^2 dx + \frac{4EA\pi^3 \xi_u \xi_v}{L^3} \int_{-L/2}^{L/2} \cos\frac{4\pi x}{L} \sin^2 \frac{2\pi x}{L} dx$$

$$+ \frac{EA\pi^4 \xi_v^3}{2L^4} \int_{-L/2}^{L/2} \sin^4 \frac{2\pi x}{L} dx + \frac{4EI\pi^4 \xi_v}{L^4} \int_{-L/2}^{L/2} \cos^2 \frac{2\pi x}{L} dx = Q_v, \tag{2.26}$$

$$m\ddot{\xi}_u \int_{-L/2}^{L/2} \sin^2 \frac{4\pi x}{L} + \frac{16EA\pi^2 \xi_u}{L^2} \int_{-L/2}^{L/2} \cos^2 \frac{4\pi x}{L} dx$$

$$+ \frac{2EA\pi^3 \xi_v^2}{L^3} \int_{-L/2}^{L/2} \cos\frac{4\pi x}{L} \sin^2 \frac{2\pi x}{L} dx = Q_u. \tag{2.27}$$

The definite integrals may be evaluated and the resulting equations become

$$\frac{3mL}{8}\ddot{\xi}_v - \frac{\pi^3 EA}{L^2}\xi_u \xi_v + \frac{3\pi^4 EA}{16L^3}\xi_v^3 + \frac{2\pi^4 EI}{L^3}\xi_v = Q_v, \tag{2.28}$$

$$\frac{mL}{8}\ddot{\xi}_u + \frac{8\pi^2 EA}{L}\xi_u - \frac{\pi^3 EA}{2L^2}\xi_v^2 = -Q_u. \tag{2.29}$$

The u displacement is usually much less than the v component and the inertial term, $\ddot{\xi}_u$, may be neglected. If there is no forcing term, Q_u, then the system reduces to a single differential equation for the v displacement,

$$\ddot{\xi}_v + a\xi_v + b\xi_v^3 = Q_v, \tag{2.30}$$

where

$$a = \frac{16\pi^4 EI}{3mL^4}, \tag{2.31}$$

$$b = \frac{\pi^4 EA}{3mL^4}. \tag{2.32}$$

Solutions to this differential equation are given in Chapter 6, Examples 6.2, 6.3, and 6.4.

2.2.8 Vector Notation and Operations

In many models, the governing equations are systems of equations with very similar forms. A powerful notation for collapsing the system of equations to a single line was originally invented by Einstein[6] as a shorthand notation. This allowed him to greatly reduce the amount of writing. In this compact form, it was also possible for him to recognize and form more abstract models. This notation is termed indicial or tensor notation.

This section does not deal exclusively with indicial notation. It also includes vector notation and how the two are related. Both methods increase the abstraction of the mathematics, thereby reducing their complexity and bulk, while retaining or even improving on one's understanding.

2.2.8.1 Elementary Operations

It is assumed that the reader is familiar with the basics of vector notation. These include the following operations:

$\vec{a} \pm \vec{b},$

$\vec{a} \cdot \vec{b},$

$\vec{a} \times \vec{b},$

$\dfrac{d}{dt}(\vec{a} \cdot \vec{b}).$

2.2.8.2 $\vec{\nabla}$ Operators

It is assumed that the reader is familiar with the grad, div, and curl vector operators. Their notation is included here for completeness. The general definition of the $\vec{\nabla}$ operators in vector notation is independent of the coordinate system used. Using the notation shown in Fig. 2.10, the grad operator acting on a scalar ϕ is defined as

$$\operatorname{grad} \phi = \lim_{\Delta V \to 0} \left[\frac{1}{\Delta V} \oiint_s \phi(\vec{n} dA) \right] = \vec{\nabla} \phi. \tag{2.33}$$

The divergence is defined as

$$\operatorname{div} \vec{a} = \lim_{\Delta V \to 0} \left[\frac{1}{\Delta V} \oiint_s \vec{a} \cdot \vec{dA} \right] = \vec{\nabla} \cdot \vec{a}, \tag{2.34}$$

---

[6] Albert Einstein (1879–1955).

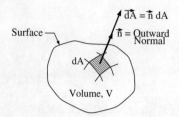

Figure 2.10. Outward area normal vector definition.

and the curl is

$$\operatorname{curl} \vec{a} = \lim_{\Delta V \to 0} \left[\frac{1}{\Delta V} \oiint_S \vec{a} \times \vec{dA} \right] = \vec{\nabla} \times \vec{a}. \tag{2.35}$$

2.2.8.3 Divergence Theorem

The Gauss[7] divergence theorem relates volume integrals to surface integrals. It is given as

$$\iiint_V (\vec{\nabla} \cdot \vec{a}) \, dV = \oiint_S \vec{a} \cdot \vec{dA}, \tag{2.36}$$

where the vectorial differential area is defined along an outward directed normal (see Fig. 2.10) as

$$\vec{dA} = \vec{n} \, dA.$$

The divergence theorem is restricted to cases where $\vec{a}$ and its partial derivatives are continuous in V and on S. The region V need not be simply connected, so long as its complete boundary is taken into account (both inner and outer). A simply connected region is one in which every closed curve can be shrunk to a point without crossing the surface boundary. Multiply connected regions are those that do not exhibit this property.

In Cartesian coordinates, Eq. (2.36) may be written as

$$\iiint_V \left(\frac{\partial a_x}{\partial x} + \frac{\partial a_y}{\partial y} + \frac{\partial a_z}{\partial z} \right) dx \, dy \, dz = \oiint_S (a_x \, dy \, dz + a_y \, dx \, dz + a_z \, dx \, dy).$$

2.2.8.4 Stokes' Theorem

Stokes'[8] theorem relates surface integrals to line integrals. It is often used in fluid mechanics, for example, to relate circulation and rotation. The theorem states that

$$\oiint_S (\vec{\nabla} \times \vec{a}) \cdot \vec{a} \, dS = \int_C \vec{a} \cdot \vec{dC}.$$

For this relationship to hold, S must be contained in a simple region where $\vec{a}$ is continuously differentiable. The path integral along C must be on a closed curve on the surface S.

---

[7] Carl Friedrich Gauss (1777–1855).
[8] George Gabriel Stokes (1819–1903).

2.2. Modeling Basics

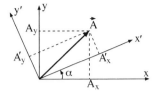

Figure 2.11. Simple rotation transformation.

2.2.8.5 Fundamental Vector Properties

Definition

A vector, by definition, is unchanged if shifted in space without change to either length or direction.

Transformation Properties

A two-dimensional vector in x-y Cartesian coordinates is shown in Fig. 2.11. If the vector $\vec{A}$ is to be transformed from the x-y coordinates to a new coordinate system x'-y', it can be shown that

$$A_x = A'_x \cos \alpha - A'_y \sin \alpha,$$
$$A_y = A'_x \sin \alpha + A'_y \cos \alpha.$$

We now introduce the definition of a *direction cosine*, which is the cosine of the smallest angle between the axes. Thus,

$$\cos(y', x) \doteq -\sin \alpha = \cos\left(\alpha + \frac{\pi}{2}\right),$$
$$\cos(x', y) \doteq \sin \alpha = \cos\left(-\alpha + \frac{\pi}{2}\right).$$

These are introduced and the transformation equations become

$$A_x = A'_x \cos(x', x) + A'_y \cos(y', x),$$
$$A_y = A'_x \cos(x', y) + A'_y \cos(y', y).$$

We now generalize to three-dimensional space and substitute 1, 2, 3 for x, y, z, yielding

$$A_1 = a_{11} A'_1 + a_{21} A'_2 + a_{31} A'_3,$$
$$A_2 = a_{12} A'_1 + a_{22} A'_2 + a_{32} A'_3,$$
$$A_3 = a_{13} A'_1 + a_{23} A'_2 + a_{33} A'_3,$$

where each of the a_{ij} are direction cosines between the i and j axes and the A'_i are components of the vector in the transformed (prime) coordinate system. These transformation equations may be rewritten in the shorthand indicial notation as a single equation:

$$A_i = \sum_{j=1}^{3} a_{ji} A'_j = a_{ji} A'_j.$$

One of the rules of indicial notation is that repeated indices are used to imply a summation. In this way, the overhead of explicitly writing out the summation is eliminated and the summation sign therefore can be dropped.

Principle of Equivalence

Vector transformations were developed arbitrarily as a transition from x', y', z' coordinates to x, y, z, but the reverse also must be possible. Therefore,

$$a_{ji} = \cos(x'_j, x_i),$$
$$a_{ij} = \cos(x_j, x'_i) = \cos(x'_i, x_j),$$

which results in the rule

$$A'_i = a_{ij} A_j, \tag{2.37}$$
$$A_i = a_{ji} A'_j. \tag{2.38}$$

This result will work for any type of coordinate system (rectilinear, polar, cylindrical). The result of Eqs. (2.37) and (2.38) is, in effect, an analytic definition of a vector. We thus may write that vectors are independent of the coordinate system, or

$$\vec{A} = A_i \vec{x}_i = A'_i \vec{x}'_i.$$

It now can be seen that first-order tensors and vectors are equivalent concepts.

Concept and Definition of a Tensor

Elements with a single subscript are termed first-order tensors. Examples include u_i, v_k, and x_j. Let us combine the definitions of these to achieve a definition for higher-order operations. Consider

$$u_i = a_{ki} u'_k,$$
$$v_j = a_{mj} v'_m.$$

Thus,

$$u_i v_j = \left[\sum_{k=1}^{3} a_{ki} u'_k\right] \left[\sum_{m=1}^{3} a_{mi} v'_k\right]$$
$$= \sum_{k=1}^{3} \sum_{m=1}^{3} a_{ki} a_{mj} u'_k v'_m,$$

or, using indicial notation rules,

$$u_i v_j = a_{ki} a_{mj} u'_k v'_m;$$

also,

$$u'_i v'_j = a_{ik} a_{jm} u_k v_m.$$

These terms have two free indices (i, j), and this defines a second-order tensor. It is possible to extend this notation and transform tensors in the same manner as vectors. This is demonstrated by the equations below:

$$T_{ij} = a_{ki} a_{mj} T'_{km},$$
$$T'_{ij} = a_{ik} a_{jm} T_{km}.$$

2.2. Modeling Basics

To determine if an entity is a tensor, we transform it and see if it behaves like a tensor. In this manner, if

$$T_{ij} = \begin{bmatrix} T_{11} & T_{12} & T_{13} \\ T_{21} & T_{22} & T_{23} \\ T_{31} & T_{32} & T_{33} \end{bmatrix}$$

is transformed, it generally behaves as an array, not necessarily as a tensor.

2.2.8.6 Introduction to Tensor Notation

This notation is used frequently in the literature. It is also useful as a shorthand notation and is a natural language for problems related to field theory. In this notation, the number of components, C, of an Nth-order tensor is

$$C = 3^N.$$

A first-order tensor ($N = 1$), also called a vector, therefore has 3 components, whereas a third-order tensor ($N = 3$) has 27 components. Many tensors also carry dummy indices, or repeated indices, which are not counted in determining the order of the tensor. In these cases, each component will have a number of terms. The number of terms, T, per component having D dummy indices is therefore

$$T = 3^D.$$

A special feature of the use of tensor notation is that repeated indices indicate summation. Consider the zeroth-order tensor

$$a_i x_i = a_1 x_1 + a_2 x_2 + a_3 x_3.$$

This has the exact character of the dot product of two vectors, $\vec{a} \cdot \vec{x}$. Note that the repeated index, i, is a dummy index and does not contribute to a computation of the order of the tensor. A first-order tensor is shown below along with some other representations. Note that, because there is a single dummy index, i, there are three components per term:

$$a_i x_{ij} = a_1 x_{1j} + a_2 x_{2j} + a_3 x_{3j}$$

$$= a_1 \begin{Bmatrix} x_{11} \\ x_{12} \\ x_{13} \end{Bmatrix} + a_2 \begin{Bmatrix} x_{21} \\ x_{22} \\ x_{23} \end{Bmatrix} + a_3 \begin{Bmatrix} x_{31} \\ x_{32} \\ x_{33} \end{Bmatrix}$$

$$= \begin{Bmatrix} [a_1 x_{11} + a_2 x_{21} + a_3 x_{31}] \\ [a_1 x_{12} + a_2 x_{22} + a_3 x_{32}] \\ [a_1 x_{13} + a_2 x_{23} + a_3 x_{33}] \end{Bmatrix}$$

$$\neq \begin{Bmatrix} a_1 x_{11} & a_1 x_{12} & a_1 x_{13} \\ a_2 x_{21} & a_2 x_{22} & a_2 x_{23} \\ a_3 x_{31} & a_3 x_{32} & a_3 x_{33} \end{Bmatrix}.$$

2.2.8.7 Tensor Algebra

A rule of tensor notation is that no index may appear more than twice in any one term and every term in a tensor equation must have the same order and number of free indices. The notation for tensor operations are

Addition. Only applies to tensors of the same order,

$$a_{ijk} + b_{ijk} = c_{ijk}.$$

Multiplication. This is a definition of an outer multiplication operation. The resulting tensor product has an order that is the sum of orders of the factors,

$$a_{ij}b_k = c_{ijk}.$$

Contraction. This is a definition of a reduction operation caused by setting two indices to be equal. Consider the second-order tensor equation

$$a_i b_j = c_{ij} = \begin{cases} (a_1 b_1 = c_{11}) & (a_1 b_2 = c_{12}) & (a_1 b_3 = c_{13}) \\ (a_2 b_1 = c_{21}) & (a_2 b_2 = c_{22}) & (a_2 b_3 = c_{23}) \\ (a_3 b_1 = c_{31}) & (a_3 b_2 = c_{32}) & (a_3 b_3 = c_{33}) \end{cases}.$$

For the special case of $i = j$, the diagonal elements of the above tensor are addressed. For this case,

$$a_i b_i = c_{ii} = a_1 b_1 + a_2 b_2 + a_3 b_3 = c_{11} + c_{22} + c_{33}.$$

2.2.8.8 Properties of Second-Order Tensors

Consider the position vector given as

$$\vec{r} = x_i = \begin{Bmatrix} x_1 \\ x_2 \\ x_3 \end{Bmatrix} = \begin{Bmatrix} x'_1 \\ x'_2 \\ x'_3 \end{Bmatrix},$$

where

$$x_i = a_{ki} x'_k,$$
$$x'_k = a_{km} x_m.$$

Differentiation yields

$$\frac{\partial x_i}{\partial x_j} = a_{ki} \frac{\partial x'_k}{\partial x_j} = a_{ki} a_{km} \frac{\partial x_m}{\partial x_j}.$$

Here, we introduce the definition of the Kronecker delta function,

$$\delta_{ij} \doteq \frac{\partial x_i}{\partial x_j} = \begin{cases} 1 & \text{for } i = j \\ 0 & \text{for } i \neq j \end{cases}.$$

This second-order tensor can be transformed as any other, yielding

$$\delta_{ij} = a_{ki}(a_{km} \delta_{mj}) = a_{ki} a_{kj},$$

2.2. Modeling Basics

and this allows us to write that

$$\delta_{ij} = a_{ki}a_{kj},$$
$$\delta'_{ij} = a_{ik}a_{jk}.$$

There is an important observation, however, that $\delta_{ij} \equiv \delta_{ji}$ and thus the elements of δ_{ij} are totally independent of the coordinate system chosen. This is termed an *invariant tensor*. Examining the expansion of δ, the first set of terms for $i = j$ are called the normalizing conditions,

$$\left.\begin{array}{l}\delta_{11} = a_{11}^2 + a_{21}^2 + a_{31}^2 = 1 \\ \delta_{22} = a_{12}^2 + a_{22}^2 + a_{32}^2 = 1 \\ \delta_{33} = a_{13}^2 + a_{23}^2 + a_{33}^2 = 1\end{array}\right\}, \tag{2.39}$$

and the others are termed orthogonality conditions,

$$\left.\begin{array}{l}\delta_{23} = \delta_{32} = a_{12}a_{13} + a_{22}a_{23} + a_{32}a_{33} = 0 \\ \delta_{13} = \delta_{31} = a_{13}a_{11} + a_{23}a_{21} + a_{33}a_{31} = 0 \\ \delta_{12} = \delta_{21} = a_{11}a_{12} + a_{21}a_{22} + a_{31}a_{32} = 0\end{array}\right\}. \tag{2.40}$$

The contraction of a second-order tensor of the form

$$T_{ij} = a_{ki}a_{mj}T'_{km}$$

with respect to the indices i and j yields

$$T_{ii} = a_{ki}a_{mi}T'_{km} = \delta_{km}T'_{km},$$

but this yields

$$T_{ii} = T'_{kk} = T'_{mm}.$$

Thus, any tensor such as T_{ii} becomes invariant with respect to the indices of contraction.

Any tensor of order greater than one that remains unaltered when any two indices are interchanged is termed a *symmetric tensor* with respect to those two indices. If all of the indices may be interchanged readily without altering the result, the tensor is *completely symmetric*. Thus,

$$c_{ijk} = c_{ikj} = c_{jki},$$
$$c_{ij} = c_{ji},$$
$$\delta_{ij} = \delta_{ji} = \begin{Bmatrix} 1 & 0 & 0 \\ 0 & 1 & 0 \\ 0 & 0 & 1 \end{Bmatrix}$$

are all examples of symmetric tensors. The tensor δ_{ij} is both symmetric and invariant.

Any tensor of order greater than one having components that only change sign when indices are interchanged is termed *antisymmetric* with respect to those two indices. If this property is exhibited for any combination of interchanged indices, the tensor is labeled *completely antisymmetric*.

Some other interesting properties are that if $b_{ij} \neq \pm b_{ji}$, then b_{ij} is neither antisymmetric or symmetric. The combination $(b_{ij} + b_{ji})$ is symmetric and $(b_{ij} - b_{ji})$ is antisymmetric. Thus, any tensor can be broken into a combination of symmetric and antisymmetric tensors, or

$$b_{ij} = \tfrac{1}{2}(b_{ij} + b_{ji}) + \tfrac{1}{2}(b_{ij} - b_{ji}).$$

2.2.8.9 The Alternating Tensor

A third-order, completely general, antisymmetric tensor, a_{ijk}, can have only three distinct values:

0	when any *two* indices are equal;
$+a_{123}$	for even permutations of 1, 2, 3;
$-a_{123}$	for odd permutations of 1, 2, 3.

The terms *even* and *odd* refer to the cyclic permutations of the indices. Thus, the even permutations are 123, 231, 312 and all others are defined as odd. The orthonormal form of such a tensor is termed the *alternating tensor* and has the properties

$$\epsilon_{ijk} \doteq \begin{cases} 0\text{---equal indices} \\ 1\text{---even permutations of 1, 2, 3.} \\ -1\text{---odd permutations of 1, 2, 3} \end{cases}$$

This tensor has a property that is useful to us in forming vector cross products. Consider

$$c_i = \epsilon_{ijk} A_j B_k, \qquad (2.41)$$

where A_j and B_k are vectors. A general transformation of these vectors may be written as

$$A_j = a_{mj} A'_m,$$
$$B_k = a_{nk} B'_n,$$

resulting in

$$c_i = a_{mj} a_{nk} \epsilon_{ijk} A'_m B'_n.$$

Multiplying both sides by a_{si} produces a transformed vector,

$$c'_s = a_{si} c_i = a_{si} a_{mj} a_{nk} \epsilon_{ijk} A'_m B'_n$$
$$= \epsilon'_{smn} A'_j B'_k = \epsilon_{smn} A'_m B'_n \qquad (2.42)$$

Thus, as demonstrated in Eqs. (2.41) and (2.42), the transformation of the product of two vectors $\vec{A}$ and $\vec{B}$ results in a vector $\vec{c} = c_i$. To demonstrate the equivalence, consider the Cartesian representation of a cross product:

$$c_i = \epsilon_{ijk} A_j B_k = \begin{vmatrix} \vec{\epsilon}_1 & \vec{\epsilon}_2 & \vec{\epsilon}_3 \\ A_1 & A_2 & A_3 \\ B_1 & B_2 & B_3 \end{vmatrix}$$
$$= (A_2 B_2 - A_3 B_2)\vec{\epsilon}_1 + (A_3 B_1 - A_1 B_3)\vec{\epsilon}_2 + (A_1 B_2 - A_2 B_1)\vec{\epsilon}_3. \qquad (2.43)$$

The reader should verify that the individual terms for C_i derived from the tensor notation are equivalent to those given in Eq. (2.43).

2.2. Modeling Basics

2.2.8.10 Derivatives of Tensors

Consider a scalar (a zeroth-order tensor) field where the scalar is dependent on a position vector into the field, or $\phi(x_1, x_2, x_3) = \phi(x_i)$. Is the result of differentiation of this scalar a vector as suggested by its representation $\partial \phi / \partial x_i$? This can be ascertained by transforming it to see if it behaves as a vector. The position vector, x_i, must obey the rule that

$$x_i = a_{ji} x'_j.$$

The derivative of the scalar function with respect to the transformed position vector, x'_i, can be written as

$$\frac{\partial \phi}{\partial x'_j} = \frac{\partial \phi}{\partial x_i} \frac{\partial x_i}{\partial x'_j} = a_{ji} \frac{\partial \phi}{\partial x_i}. \tag{2.44}$$

One can also write the scalar as a function of the transformed position vector, $\phi = \phi(x'_i)$, so that its derivative is

$$\frac{\partial \phi}{\partial x_j} = \frac{\partial \phi}{\partial x'_i} \frac{\partial x'_i}{\partial x_j} = a_{ij} \frac{\partial \phi}{\partial x'_i}. \tag{2.45}$$

After comparison of Eqs. (2.44) and (2.45), it is obvious that $\partial \phi / \partial x_j$ behaves as a vector.

Another property of differentiation is that it results in a tensor of one order higher than its operand. Additionally, the derivative may be transformed as any other tensor. Therefore,

$$\frac{\partial A_i}{\partial x_k} = a_{mk} a_{ji} \frac{\partial A'_j}{\partial x'_m}.$$

2.2.8.11 Isotropic Tensors

An isotropic tensor is one for which all of the components are invariant. Thus, any Cartesian tensor that transforms into itself under an axis rotation is an isotropic tensor. Examples of this include both the Kronecker delta function, δ_{ij}, and the alternating tensor, ϵ_{ijk}.

The most general fourth-order isotropic tensor must be transformable and can be expressed as

$$C_{ijkm} = \frac{\partial x_i}{\partial x'_r} \frac{\partial x_j}{\partial x'_s} \frac{\partial x_k}{\partial x'_t} \frac{\partial x_m}{\partial x'_u} C'_{rstu} = a_{ri} a_{sj} a_{tk} a_{um} C'_{rstu}.$$

This same tensor also can be written in terms of permutations of the four indices, or

$$C_{ijkm} = \delta_{ij} \delta_{km} k_1 + \delta_{ik} \delta_{jm} k_2 + \delta_{im} \delta_{jk} k_3,$$

where k_1, k_2, and k_3 are constants.

Consider a general second-order tensor, S_{ij}, which is not isotropic. It is possible to break down this tensor into two elements as

$$S_{ij} = \left(S_{ij}\right)_{\text{isotropic}} + \left(S_{ij}\right)_{\text{deviatric}}. \tag{2.46}$$

The term *deviatric* thus refers to the difference between the general tensor and an isotropic component. We also can equate the diagonal terms of the isotropic tensor to those of the full tensor, or

$$S_{ii} = (S_{ii})_{\text{isotropic}} = S_{11} + S_{22} + S_{33} \doteq k \delta_{ii},$$

where $k = S_{11} + S_{22} + S_{33} = $ scalar constant. This, of course, requires that the sum of the diagonal terms of the deviatric are zero. Thus we write the general relationship

$$S_{ij} = k\delta_{ij} + S_{ij}^o.$$

In this equation, the term S_{ij}^o is the deviatric term in Eq. (2.46). This relationship has proven useful in expressing the stress tensor in fluid and solid mechanics. Here the diagonals of the isotropic term are each taken to be the mean thermodynamic pressure, P. Thus

$$S_{ij} = -P\delta_{ij} + S_{ij}^o.$$

2.2.8.12 Vector-Tensor Equivalence

There are equivalent operations for both vector and tensor notation. These are readily confirmed by expanding Eq. (2.35):

$$\vec{\nabla} \times \vec{A} = \vec{C}$$

$$= \begin{vmatrix} \vec{\epsilon}_1 & \vec{\epsilon}_2 & \vec{\epsilon}_3 \\ \frac{\partial}{\partial x_1} & \frac{\partial}{\partial x_2} & \frac{\partial}{\partial x_3} \\ A_1 & A_2 & A_3 \end{vmatrix}$$

$$= \left(\frac{\partial A_3}{\partial x_2} - \frac{\partial A_2}{\partial x_3}\right)\vec{\epsilon}_1 + \left(\frac{\partial A_1}{\partial x_3} - \frac{\partial A_3}{\partial x_2}\right)\vec{\epsilon}_2 + \left(\frac{\partial A_2}{\partial x_1} - \frac{\partial A_1}{\partial x_2}\right)\vec{\epsilon}_3.$$

This can be shown to be equivalent to the tensor notation

$$\vec{\nabla} \times \vec{A} = \vec{C} = \epsilon_{ijk}\frac{\partial A_k}{\partial x_j}$$

and is in agreement with the definition of the cross product given as Eq. (2.43). In a similar manner, the gradient is written

$$\vec{\nabla}\phi = \frac{\partial \phi}{\partial x_i}$$

and the dot product as

$$\vec{\nabla} \cdot \vec{A} = \frac{\partial A_i}{\partial x_i}.$$

2.3 INTEGRAL EQUATIONS

In many situations, it may prove of great value to formulate the model as an integral equation rather than as a differential equation. In this section, the conservation equations are formulated in this manner.

2.3.1 Basic Laws for Continuous Matter

In general, there are four basic laws that must be satisfied when describing a nonrelativistic system. They are

1. conservation of matter (continuity equation),
2. conservation of momentum (linear and/or angular momentum),
3. conservation of energy (first law of thermodynamics), and
4. principle of increase of entropy of the universe (second law of thermodynamics).

In addition, there are usually subsidiary equations or relations called constituitive equations which relate the properties of the matter within the system being analyzed. These are equations that describe the behavior of the material. For example, the physical properties of gases at low pressures can be interrelated by the equation known as the ideal gas law:

$$P = \rho RT. \tag{2.47}$$

In fluid mechanics, many of the basic equations assume a linear relationship between the shear stress, τ_{ij}, and the spatial gradient of the velocity, $\partial U_j/\partial x_i$. Fluids that behave in this manner are termed Newtonian and are characterized by the relationship

$$\tau_{ij} = \mu \frac{\partial U_j}{\partial x_i}. \tag{2.48}$$

Here, a shorthand notation called indicial notation has been used. As an example, to compute the shear stress τ_{xy}, we have

$$\tau_{xy} = \mu \frac{\partial U_2}{\partial x_1} = \mu \frac{\partial v}{\partial x}. \tag{2.49}$$

In solid mechanics, the same sort of relationship is used for linear materials but the stress, S, is proportional to the strain, ϵ, not the strain rate. Such materials are termed as linearly elastic and obey Hooke's law:

$$S = E\epsilon. \tag{2.50}$$

Here, the constant of proportionality, E, is the modulus of elasticity. This is an equation representing the uniaxial stress-strain behavior. A more general formulation would be written as

$$S_{ij} = E_{ijkl}\epsilon_{kl}. \tag{2.51}$$

The fourth-order tensor E_{ijkl} has 81 elements that represent material constants. For isotropic materials, these 81 elements can be reduced to only two independent constants, the modulus of elasticity and the shear modulus.

2.3.2 General Conservation Formulation

In this text, a *system analysis* is defined as one in which the matter under study is constant. Generally, we define some region in space that we wish to analyze and draw the boundaries for this system, which are depicted as the dotted lines in Fig 2.12(a). Matter may not cross the control surface in a system analysis. The term *control volume analysis* implies that attention is to be focused on a specific region in space, as shown in Fig 2.12(b). In this form of analysis, matter may cross the boundaries of the control surface, but the

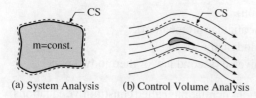

Figure 2.12. A comparison of system and control volume boundaries.

(a) System Analysis (b) Control Volume Analysis

quantity of matter and the energy it carries with it must be accounted for on the basis of conservation principles. Between two times that are infinitesimally separated, the control volume analysis and the system analysis are equal. Conservation equations, however, generally are specified for systems. Because fluid systems generally undergo continuous distortion and deformation, it is difficult to identify the same mass at different times. This provides the motivation for a formulation based on a control-volume viewpoint rather than a system viewpoint. The principal difference between the two is that in a control volume analysis, matter can cross the control surface.

In the following derivation, the quantity to be conserved is an arbitrary extensive property, N. An *extensive property* is one in which the property depends on the quantity of matter involved. Examples of extensive properties include volume, kinetic energy, and mass. An *intensive property* is one whose value is independent of the quantity of matter such as pressure, temperature, or density. Thus, if one cuts a piece of cheese in two, the temperature (an intensive property) of the two new pieces remains the same whereas the volume (an extensive property) of each piece is half of the previous measurand.

The beginning of the derivation of a general conservation equation is the definition of a derivative,

$$\left. \frac{dN}{dt} \right]_{\text{system}} = \lim_{\delta t \to 0} \frac{N_{t_0+\delta t} - N_{t_0}}{\delta t}. \qquad (2.52)$$

As shown in Fig 2.13, if we begin our analysis with a system and control volume that are identical at t_0, at some time δt later the system will have moved slightly away from the control volume since it is fixed in space. The flowfield is considered to be arbitrary and can be described using either an inertial or a noninertial reference frame. Note that region B is a part of the system that is always within the control volume. At time $t_0 + \delta t$, the system occupies both regions B and C, and at time t_0, the system and control volume coincide. If we define $\eta = dN/dm$ as N per unit mass, then we may write the total N within the system in terms of a volume integral, or

$$N_{\text{system}} = \int_{\text{system}} \eta \, dm = \iiint_{\text{system}} \eta \rho \, dV. \qquad (2.53)$$

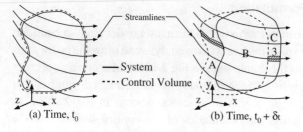

Figure 2.13. System, and control volume at two infinitesimally separate times.

2.3. Integral Equations

If we evaluate Eq. (2.53) at the two times, t_0 and $t_0+\delta t$, and substitute these into Eq. (2.52), we obtain

$$\left.\frac{dN}{dt}\right]_{\text{system}} = \lim_{\delta t \to 0} \frac{\left[\iiint_V \eta\rho\, dV\right]_{t_0+\delta t} + \left[\iiint_3 \eta\rho\, dV\right]_{t_0+\delta t} - \left[\iiint_1 \eta\rho\, dV\right]_{t_0+\delta t} - \left[\iiint_V \eta\rho\, dV\right]_{t_0}}{\delta t}. \tag{2.54}$$

Because the limit of a sum of terms is the sum of the individual limits of these terms, this can be rewritten as

$$\left.\frac{dN}{dt}\right]_{\text{system}} = \lim_{\delta t \to 0} \frac{\left[\iiint_V \eta\rho\, dV\right]_{t_0+\delta t} - \left[\iiint_V \eta\rho\, dV\right]_{t_0}}{\delta t}$$
$$+ \lim_{\delta t \to 0} \frac{\left[\iiint_C \eta\rho\, dV\right]_{t_0+\delta t}}{\delta t} - \lim_{\delta t \to 0} \frac{\left[\iiint_A \eta\rho\, dV\right]_{t_0+\delta t}}{\delta t}. \tag{2.55}$$

The first term on the right-hand side of Eq. (2.55) is recognizable as

$$\frac{\partial N_{CV}}{\partial t} = \frac{\partial}{\partial t}\iiint_{CV} \eta\rho\, dV. \tag{2.56}$$

If we define $\vec{U}_r$ as the velocity of the flowfield relative to the control surface S and an outward directed normal differential area as $d\vec{S} = \vec{n}\, dS$, the second term on the right-hand side of Eq. (2.55) becomes

$$\lim_{\delta t \to 0} \frac{\left[\iiint_3 \eta\rho\, dV\right]_{t_0+\delta t}}{\delta t} = \iiint_{S_3} \eta\rho\vec{U}_r \cdot d\vec{S} \tag{2.57}$$

and the third term becomes

$$-\lim_{\delta t \to 0} \frac{\left[\iiint_1 \eta\rho\, dV\right]_{t_0+\delta t}}{\delta t} = \iiint_{S_1} \eta\rho\vec{U}_r \cdot d\vec{S}. \tag{2.58}$$

Combining Eqs. (2.57) and (2.58) yields a single integral over the entire surface. Equation (2.55) now can be rewritten by substituting Eqs. (2.56) through (2.58) and combining the integral terms to obtain

$$\left[\frac{\mathcal{D}N}{\mathcal{D}t}\right]_{\text{system}} = \left[\oiint_S \eta(\rho\vec{U}_r \cdot d\vec{S}) + \frac{\partial}{\partial t}\iiint_V \eta\rho\, dV\right]_{\substack{\text{control}\\\text{volume}}}. \tag{2.59}$$

The integral on the right side of this equation with the $\rho\vec{U}_r d\vec{S}$ argument represents the *flux* of N across the control surface. The other integral represents the temporal change in the *storage* of N within the control volume. For the first time, we are using a notation termed the substantial derivative [8], $\mathcal{D}(\ )/\mathcal{D}t$. This notation symbolizes that the derivative is one that is with respect to the moving reference frame of the system under study. The substantial derivative can be written in indicial notation as

$$\frac{\mathcal{D}(\,)}{\mathcal{D}t} = \frac{\partial(\,)}{\partial t} + u_i \frac{\partial(\,)}{\partial x_i}. \tag{2.60}$$

In summary, Eq. (2.59) is a conservation equation that relates the conservation from a systems viewpoint and a control-volume-analysis viewpoint. It describes the conservation of an arbitrary extensive property, N, crossing the control surface, S, around the control volume, V. Using words, this can be expressed as

$$\left[\begin{array}{c} \text{total rate of} \\ \text{change of } N \end{array}\right]_{\text{system}} = \left[\left(\begin{array}{c} \text{net rate of flux} \\ \text{loss of } N \text{ across } S \end{array}\right) + \left(\begin{array}{c} \text{rate of gain of} \\ N \text{ within } V \end{array}\right)\right]_{\substack{\text{control} \\ \text{volume}}}. \tag{2.61}$$

2.3.2.1 Conservation of Matter

As an application of the general conservation theme of Eq. (2.59), let us consider the case in which the arbitrary extensive property is $N = mass$, and thus $\eta = dN/dm = 1$. The mass of a system is, by definition, constant and the application of Eq. (2.59) yields

$$\left[\frac{\mathcal{D}m}{\mathcal{D}t}\right]_{\text{system}} = 0 = \left[\oiint_S (\rho \vec{U}_r \cdot d\vec{S}) + \frac{\partial}{\partial t} \iiint_V \rho \, dV\right]_{\substack{\text{control} \\ \text{volume}}}. \tag{2.62}$$

In this result, the first integral on the right side of the equation represents the net flux of matter across the control surface. This must be counterbalanced exactly by the rate of accumulation or loss of matter within the control volume. This equation often is termed the continuity equation. It may not be recognized immediately as such because of the integral formulation. Leibnitz's[9] rule is stated as follows [54]:

If $Y(t) = \int_q^r f(x,t)\,dx$, where q and r are differentiable functions of t and where $f(x,t)$ and its derivative are continuous in x and t, then

$$\frac{dY}{dt} = \int_q^r \frac{\partial f}{\partial t} dx + f[q(t),t]\frac{dq}{dt} - f[r(t),t]\frac{dr}{dt}. \tag{2.63}$$

If the control volume, V, is assumed to be time invariant, the application of Leibnitz's rule to Eq. (2.62) may be shown to yield the more common form

$$\frac{\partial \rho}{\partial t} + \frac{\partial}{\partial x_i}(\rho U_i) = 0. \tag{2.64}$$

For incompressible fluids, this relationship can be simplified further to

$$\frac{\partial U_i}{\partial x_i} = 0 = \frac{\partial u}{\partial x} + \frac{\partial v}{\partial y} + \frac{\partial w}{\partial z}. \tag{2.65}$$

---

[9] Gottfried Wilheim Leibnitz (1646–1716).

2.3.2.2 Conservation of Momentum

We next derive an equation for the conservation of momentum. Here we let our arbitrary extensive property, N, become a vector, $\vec{N}$, equal to the momentum of a quantity of matter of mass m and velocity $\vec{U}$:

$$\vec{N} = m\vec{U}, \tag{2.66}$$

and thus the intensive form of $\vec{N}$ is

$$\vec{\eta} = \frac{d\vec{N}}{dm} = \vec{U}. \tag{2.67}$$

Equation (2.59) therefore becomes

$$\left[\frac{\mathcal{D}(m\vec{U})}{\mathcal{D}t}\right]_{\text{sys}} = \sum \vec{F}_{\text{sys}} = \left[\oiint_S \vec{U}(\rho \vec{U}_r \cdot d\vec{S}) + \frac{\partial}{\partial t}\iiint_V \vec{U}\rho\,dV\right]_{\text{control volume}}. \tag{2.68}$$

We commonly break the forces into two groups, surface and body forces:

$$\vec{F} = \vec{F}_s + \vec{F}_b. \tag{2.69}$$

An additional problem arises because the velocity term in the term $(\rho \vec{U}_r \cdot d\vec{S})$ must be relative to the control surface, whereas the other velocity term, $\vec{U}$, must be an absolute velocity expressed in absolute, or "world" coordinates.

The control volume chosen will dramatically affect the analysis of this problem. Through the careful selection of the control surface, it may be possible to simplify the analysis. This is perhaps best demonstrated through the use of an example.

EXAMPLE 2.4: ROCKET CAR

Consider the rocket car shown in Fig. 2.14. If the car is at steady state, what is the net drag, D? It is readily seen that we do not have sufficient information to specify the velocity around the entire surface for CV_1. However, by choosing the control surface CV_2, most of the surface is solid and *no matter crosses the control surface*. Therefore, $\vec{U}_r \cdot d\vec{A} = 0$ over most of the surface.

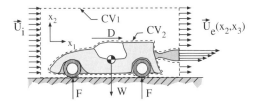

Figure 2.14. Simple rocket car on wheels.

From experimental data, suppose that the following values characterize the velocities, $\vec{U}$, exit area, $\vec{S}_e$, and density, ρ:

$$\begin{aligned}
\vec{U}_i &= U_0 \vec{e}_1, \\
\vec{U}_e &= (U_0 + U_1)\vec{e}_1 + U_2 \vec{e}_2, \\
\vec{S}_e &= hw\vec{e}_1, \\
U_0 &= 100 \text{ ft/s}, \\
U_1 &= 1500 \text{ ft/s}, \\
U_2 &= (100 \text{ s}^{-1})x_2, \quad 0 \le x_2 \le 3 \text{ ft}, \\
\rho &= 0.05 \text{ lbm/ft}^3, \\
h &= 3.0 \text{ ft}, \\
w &= 3.0 \text{ ft}.
\end{aligned} \quad (2.70)$$

The resulting momentum equation becomes

$$D\vec{e}_1 + (2F - W)\vec{e}_2 = \left[\oiint_S \vec{U}(\rho \vec{U} \cdot d\vec{S}) + \frac{\partial}{\partial t} \iiint_V \vec{U} \rho \, dV \right]_{\text{control volume}}. \quad (2.71)$$

Since the problem is only to address the steady-state condition, the last term on the right side of this equation is zero. This can be true only if the rate of change of the mass of the rocket car is small over the period of the analysis. In addition, except across the nozzle exit, the term $\vec{U} \cdot d\vec{A} = 0$. The area integral thus becomes

$$\oiint_S \vec{U}(\rho \vec{U}_r \cdot d\vec{S}) = \oiint_e \vec{U}_e(\rho(\vec{U}_e - \vec{U}_i) \cdot d\vec{S}_e) \quad (2.72)$$

$$= \int_{x_2=0}^{h} \int_{x_3=0}^{w} [(U_0 + U_1)\vec{e}_1 + U_2 x_2 \vec{e}_2)]\rho[(U_1 + U_0)\vec{e}_1 + U_2 x_2 \vec{e}_2 - U_0 \vec{e}_1 \cdot (dx_2 dx_3 \vec{e}_1) \quad (2.73)$$

$$= \int_{x_2=0}^{h} \int_{x_3=0}^{w} [(U_0 + U_1)\vec{e}_1 + U_2 x_2 \vec{e}_2][\rho(U_1)(dx_2 dx_3)]. \quad (2.74)$$

When inserted into the momentum equation along with the constant values, this becomes

$$D = 3.15 \times 10^4 \text{ lbf}, \quad (2.75)$$
$$2F - W = 3.15 \times 10^3 \text{ lbf}. \quad (2.76)$$

2.3.3 First Law of Thermodynamics

The first law is a statement of a principle based on the assumption that energy is conserved on a macroscopic scale. Therefore, we take energy as our arbitrary extensive property, or $N = E$. If we take a fixed quantity of matter, then the change in energy of our system is equivalent to the transfer of energy at the boundary, whether in the form or work, W, or heat transfer, Q. That is,

$$\frac{\mathcal{D}E}{\mathcal{D}t} = \frac{\delta Q}{dt} - \frac{\delta W}{dt}. \quad (2.77)$$

2.4. Differential Equations

The negative sign on the work term is due to the classical convention which states that the work is assigned a positive sign when a system performs useful work. With respect to energy transfer, this represents a loss of energy to the system. Similarly, heat transfer to a system is considered positive. Considering the matter within the control volume, we may characterize its total energy in words as

$$E = \text{kinetic} + \text{potential} + \text{internal} + \cdots \tag{2.78}$$

or, on a per-unit-mass basis, these factors may be written mathematically as

$$e = \frac{dN}{dm} = \frac{V^2}{2} + gz + u + \cdots. \tag{2.79}$$

Inserting this information into Eq. (2.59) yields

$$\frac{\delta Q}{\delta t} - \frac{\delta W}{\delta t} = \oiint_S e(\rho \vec{U}_r \cdot d\vec{S}) + \frac{\partial}{\partial t} \iiint_V e\rho\, dV. \tag{2.80}$$

Here, the symbol $\delta(\ )/\delta t$ is used to represent functions that are path dependent, not merely state dependent. Interestingly, this is not the usual form for the first law because the work term includes the usual shaft and electrical forms of work as well as work done in moving the control volume against the surroundings. Therefore, we break the work into two components, a point-transfer term and a term due to flow at a pressure P across the boundary:

$$\frac{\delta W}{dt} = \frac{\delta W}{\delta t}\bigg|_{\text{point}} + \frac{\delta W}{\delta t}\bigg|_{\text{flow}} \tag{2.81}$$

$$= \frac{\delta W}{\delta t}\bigg|_{\text{point}} + \oiint_S \frac{P}{\rho}(\rho \vec{U}_r \cdot d\vec{S}). \tag{2.82}$$

Combining this result with that of Eqs. (2.80) and (2.79) and gathering terms yields the classical form of the first law of thermodynamics:

$$\frac{\delta Q}{dt} - \frac{\delta W}{dt} = \oiint_S \left(h + \frac{V^2}{2} + gz\right)(\rho \vec{U}_r \cdot d\vec{S}) + \frac{\partial}{\partial t}\iiint_V \left(u + \frac{V^2}{2} + gz\right)\rho\, dV. \tag{2.83}$$

Here the point subscript on the work term has been dropped and the enthalpy has been introduced ($h = u + pv$).

2.4 DIFFERENTIAL EQUATIONS

There are many names or labels placed on differential equations. Mathematicians use these labels to infer something of the properties of the systems being described by the equations. These include labels such as *ordinary differential equations*, *partial differential equations*, *homogeneous differential equations*, *nonhomogeneous differential equations*, *differential equations with constant coefficients*, and of course *linear* and *nonlinear differential equations*. Because of the importance of these classifications to mathematical modeling, it is appropriate to consider the types of systems we will encounter and discuss their general properties and behavior.

It is assumed that the reader is familiar with ordinary and partial differential equations and the usual solution techniques. Most classical treatments also note that the usual form of differential equations is

$$G[U(t)] = F(t). \tag{2.84}$$

Here, the representation $G[\ ]$ is a general operator and typically includes derivatives, powers, and other linear or nonlinear functions. The variable $F(t)$ is called the *forcing function* because it forces the dependent variable U to change with respect to the independent variable t. The initial or boundary conditions also may be used to promote change in the variable U. The function G also may contain parameters that are time dependent. A third means of promoting change in the dependent variable is through these parameters. Although Eq. (2.84) is shown as an ordinary differential equation, the same classification is valid for a system of partial differential equations.

2.4.1 Ordinary Differential Equations

2.4.1.1 Linear Differential Equations

These are equations of the general form

$$A_n(x)\frac{d^n y}{dx^n} + A_{n-1}(x)\frac{d^{n-1} y}{dx^{n-1}} + \cdots + A_1(x)\frac{dy}{dx} + A_0(x)y = F(x), \tag{2.85}$$

where x is the independent variable, y is the dependent variable, and A, F are functions of the independent variable. Although this has been written as a single equation, the same rule applies for systems of equations. Although most of the methods depicted in this text are developed for equations having a single independent variable, many can be readily expanded to higher-order systems of equations.

2.4.1.2 Nonlinear Differential Equations

These are many differential equations that may not be put into the linear form described in Eq. (2.85). Examples of both linear and nonlinear differential equations include the following:

$$\frac{dy}{dx} + y = x^2 \qquad \text{Linear,}$$

$$x\frac{dy}{dx} + y = \sin(x) \qquad \text{Linear,}$$

$$\left[\frac{dy}{dx}\right]^2 + y = x^2 \qquad \text{Nonlinear,}$$

$$y\frac{dy}{dx} + y = \sin(x) \qquad \text{Nonlinear,}$$

$$\frac{dy}{dx} + e^y = x^3 \qquad \text{Nonlinear.}$$

The definition of the term nonlinear, therefore, specifically means that the nonlinearity is with respect to the dependent variable, not the independent variable.

2.4.2 Partial Differential Equations

Consider the partial differential equation describing a temperature field that is both spatially and temporally varying:

$$\rho C_p \frac{\partial u}{\partial t} = \frac{\partial}{\partial x}\left(k \frac{\partial u}{\partial x}\right) \quad (2.86)$$

where ρ = mass density, C_p = specific heat, k = thermal conductivity, u = temperature (dependent) variable, t = time (independent) variable, and x = spatial (independent) variable. This equation is linear if ρ, C_p, and k are constants or functions of either independent variable, and x or t. However, it is considered nonlinear if any of these are functions of the dependent variable, the temperature, u.

2.5 EXAMPLES OF NONLINEAR SYSTEMS

There are many examples of nonlinear systems in natural processes. Fortunately, many of these systems are not so nonlinear that they cannot be approximated using linear techniques for many useful design studies. The following sections list some of the useful modeling equations for various fields to demonstrate the wide application of nonlinear modeling.

2.5.1 Heat Transfer

Consider the effect of electromagnetic energy being radiated and absorbed at the surface of a thermal capacitance. The equation describing the variation of the temperature of this object is

$$C \frac{dT}{dt} = k\left[T_s^4 - T^4\right] \quad (2.87)$$

where C = thermal capacitance, T = temperature of capacitance, T_s = temperature of source, k = constant, and t = time. In this equation, the nonlinearity arises from the T^4 terms in the differential equation. Any temperature dependence of the object's thermal capacitance or thermal conductivity is also a nonlinear effect.

2.5.2 Thermodynamics

One form of Euler's[10] equation [20] is a well-known differential equation that mathematically models a spherical cloud of gas. The differential equation describing the variation of gravitational potential, ϕ, with respect to distance r from the center of the cloud is

$$\frac{d^2\phi}{dr^2} + \frac{2}{r}\frac{d\phi}{dr} + \phi^n = 0, \quad (2.88)$$

---

[10] Leonhard Euler (1707–1783).

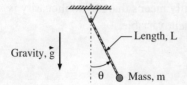

Figure 2.15. Simple pendulum in a gravitational field.

where $\phi(r)=$ gravitational potential at a point in a spherical cloud of gas; $r =$ radial distance from center of cloud; and $n = 0,1$ produces a linear equation, and other values yield a nonlinear system.

2.5.3 Mechanics

One of the classical problems in mechanics is that of a pendulum in a gravitational field. This simple system is shown in Fig. 2.15. Galileo[11] often is noted as having discovered the principle of the pendulum and is certainly the first to have stated its behavior mathematically. Although Yûnis[12] [43] observed the regularity in the period of the pendulum in the twelfth century, Galileo correctly stated that the period of a pendulum of length L was proportional to $\sqrt{L}$ in 1637, provided that the amplitude of oscillation was sufficiently small. The differential equation describing the motion of the pendulum is given as

$$ml^2 \frac{d^2\theta}{dt^2} = -mgl \sin\theta, \tag{2.89}$$

where $m =$ point mass on pendulum, $g =$ local acceleration of gravity, $l =$ pendulum length, $\theta =$ angle from vertical, and $t =$ time.

For small angles, this nonlinear equation may be linearized if one assumes $\sin\theta \approx \theta$, thus reducing the governing equation to

$$\frac{d^2\theta}{dt^2} + \frac{g}{l}\theta = 0. \tag{2.90}$$

The assumption of small angles restricts the formulation derived above to angles below approximately 30 deg to keep the errors in $\sin\theta \approx \theta$ below 10%.

2.5.4 Simple Suspension System

For one-dimensional motion, x, of a suspended mass, m, under the action of an external force, F, with a damper having a damping coefficient, C, and spring with constant k, it may be shown that

$$m\frac{d^2x}{dt^2} = F - F_{\text{spring}} - F_{\text{damper}}. \tag{2.91}$$

The damper force may be written as $f_1(dx/dt)$, and the spring force as $f_0(x)$. Thus,

$$m\frac{d^2x}{dt^2} + f_1\left(\frac{dx}{dt}\right) + f_0(x) = F. \tag{2.92}$$

---

[11] Galileo Galilei (1564–1642).
[12] Ibn Yûnis the Younger (ca. 1200).

2.5. Examples of Nonlinear Systems

It is not possible to determine by inspection whether this equation is linear or nonlinear because this requires knowledge of the character of the functions f_0 and f_1. The equation can only be linear, however, when $f_1(dx/dt) = A_1(t)\, dx/dt$ and $f_0(x) = A_0(t)x$.

2.5.5 Fluid Mechanics

The following are the two-dimensional (x_1, x_2) forms of the Navier[13]-Stokes equations [41] from fluid mechanics:

x_1-dir

$$\frac{\partial u}{\partial t} + \left\{ u \frac{\partial u}{\partial x_1} + v \frac{\partial u}{\partial x_2} \right\} - \nu \left[\frac{\partial^2 u}{\partial x_1^2} + \frac{\partial^2 u}{\partial x_2^2} \right] = X_1 - \frac{1}{\rho} \frac{\partial p}{\partial x_1}, \qquad (2.93)$$

x_2-dir

$$\frac{\partial v}{\partial t} + \left\{ u \frac{\partial v}{\partial x_1} + v \frac{\partial v}{\partial x_2} \right\} - \nu \left[\frac{\partial^2 v}{\partial x_1^2} + \frac{\partial^2 v}{\partial x_2^2} \right] = X_2 - \frac{1}{\rho} \frac{\partial p}{\partial x_2}, \qquad (2.94)$$

where u, v = velocity components in x_1, x_2 directions; X_1, X_2 = body forces in the x_1, x_2 directions; p = thermodynamic pressure; ρ = mass density; ν = kinematic viscosity; x_1, x_2 = Cartesian coordinate values; and t = temporal coordinate value.

These are coupled nonlinear partial differential equations describing the fluid flowfield in terms of the x_1, x_2 velocity components $u(x_1, x_2, t)$ and $v(x_1, x_2, t)$, respectively. The nonlinearity of these equations results from the *convective terms* enclosed in curly braces. The forcing function is the entire right side of the equation and has components due to any conservative force fields acting on the fluid and a pressure gradient term.

2.5.6 Continuity Equation

The equation describing the conservation of mass within a fluid field are given by the following equation:

$$\frac{\partial \rho}{\partial t} = \frac{\partial \rho u}{\partial x_1} + \frac{\partial \rho v}{\partial x_2}. \qquad (2.95)$$

It is seen that this equation must be solved in concert with that describing the velocity field. Since both u and v are functions of x_1, x_2, and t, the resulting differential equation for ρ is nonlinear.

2.5.7 Blasius Equation

The equation describing steady flow of an incompressible fluid over a flat plate is derived and analyzed by Schlichting [41]. The boundary-layer equations may be transformed using a class of similarity transformations (see Section 2.7) and are reduced to

$$\frac{d^3 f(\eta)}{d\eta^3} + f(\eta) \frac{d^2 f(\eta)}{d\eta^2} = 0, \qquad (2.96)$$

---

[13] Louis M.H. Navier (1785–1836).

where $f(\eta)$ is the Blasius function and η is defined as

$$\eta = \frac{x_2}{2}\left[\frac{U}{\nu x_1}\right]^{1/2}, \tag{2.97}$$

where U = freestream velocity, ν = kinematic viscosity, x_1 = streamwise distance along plate, x_2 = height above plate, and the boundary conditions are transformed to

$f(0) = 0,$
$df/d\eta(0) = 0,$
$df/d\eta(\eta) \to 2$ as $\eta \to \infty.$

This transformation and the method upon which it is based is presented in greater detail in Section 2.7.

2.5.8 Vehicle Dynamics

Consider a simplistic model of an automotive vehicle. In our example, the action of the two rear wheels is assumed to behave as if there were a single wheel on the rear axle, and the front wheels are combined similarly. The vehicle mass is assumed to behave as a point mass, albeit with an inertia I, located at the center of gravity of the vehicle, and on a massless rod connecting the front and rear axles. These simplifications are depicted schematically in Fig. 2.16. The governing differential equation for this greatly simplified system is

$$I\frac{d\omega}{dt} = aF_f - bF_r, \tag{2.98}$$

where I = yaw moment of inertia, ω = yaw velocity, F_f = lateral force at front wheels, and F_r = lateral force at rear wheels. In the above formulation, the F are forces that must be described in terms of slip angles, load, caster, camber, coefficients of friction, etc. These relationships are responsible for the nonlinearity in this problem. As an initial model, the dependence of these forces on the weight on the wheels, the contact patch size, the coefficient of friction, and many other factors is considered secondary to their dependence on the slip angle. A first approximation of this is

$$F_F = C_F \alpha_F, \qquad F_R = C_R \alpha_R. \tag{2.99}$$

This linear model is only valid for very small slip angles. As depicted in Fig. 2.17, the front and rear slip angles, α_F, α_R are seen to be dependent on the yaw velocity, ω. These

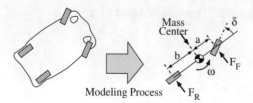

Figure 2.16. Simplified model of automotive steering problem.

2.6. Normalization of Equations

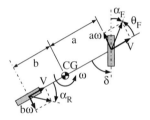

Figure 2.17. Effect of steering angle and yaw velocity on slip angle.

are written as

$$\alpha_f = \theta_F - \delta = \arctan \frac{a\omega}{V} - \delta, \quad (2.100)$$

$$\alpha_R = \arctan \frac{-b\omega}{V}, \quad (2.101)$$

yielding a simplified model

$$I\frac{d\omega}{dt} = aC_F \left(\arctan \frac{a\omega}{V} - \delta \right) - bC_R \arctan \frac{-b\omega}{V}.$$

2.6 NORMALIZATION OF EQUATIONS

Every modeling problem involves a solution space for both dependent and independent variables. When modeling the motion of light through space, for example, the distances of importance are measured in angstroms for the wavelengths of the light waves, and light-years for their motion, a disparity in scale of 10^{26}. This gross difference in scale has implications for both analytical and computational analysis of the models. What is introduced here is a method of normalizing the scales for such problems and concurrently nondimensionalizing the problem. This technique is just as important for linear systems, and often can render solutions that are parameter independent, while defining groups of terms that have physical significance.

This technique is also useful in reducing the possibility of roundoff errors in numerical solution techniques. This may be especially important whenever a set of differential equations must be solved. In such cases, the variables often differ greatly in numerical value, and this compounds the numerical difficulties. In some cases, this may be solely due to the selection of the units of the variables involved.

The goal of this method is to find a transformation of variables such that their new analogues have values that are always of the order of one. The technique employed is generally one of introducing a linear scaling value for each dependent and independent variable, transforming the governing equations into relations where all terms are approximately of unit magnitude. This operation is termed a normalization transformation. Any forcing functions for the problem also are scaled in a similar manner. The recipe for this transformation is relatively simple. It is

- Scale all dependent and independent variables using the following as a model: $T = (t - t_0)/t_f$, where t = original independent variable, t_0 = offset, t_f = characteristic value representing the dynamic range of t, T = new variable of order magnitude of 0 to 1.

- Generate all derivatives and functions.
- Substitute these into the governing equations and simplify, simplify, simplify.
- Transform all initial or boundary conditions.
- Check all terms to make sure that they are dimensionless. If they are not, you have made an error.

The governing equations now describe variables that are of unit magnitude (both dependent and independent).

The choice of scaling and offset values may not be readily obvious, but this is also an opportunity to rely on the insight or experience of the modeler. If properly chosen, all of the system's variables will be of order one and have an approximate value somewhere between $(-1, 1)$. If the dependent and independent variables exhibit changes of the order of one, then the derivatives also must be of order one. The values of the scaling constants may be chosen when the transformations are defined, or they may be chosen to simplify the differential equation or boundary conditions. This is perhaps best explained using the following examples. The first example is a simple linearized pendulum, the second the well-known linear spring-mass-dashpot system, and the third is a nonlinear electrical circuit with coupled thermal and current-flow equations.

EXAMPLE 2.5:

As described in Section 2.5.3, the linearized governing equation for a pendulum is

$$\frac{l}{g}\frac{d^2\theta}{dt^2} + \theta = 0, \qquad \theta(0) = \theta_0.$$

If we introduce the scaling transformations

$$y = \theta/\theta_{\max}, \qquad x = t/\tau,$$

the governing equation becomes

$$\frac{l}{g\tau^2}\frac{d^2y}{dx^2} + y = 0, \qquad y(0) = \theta_0/\theta_{\max}.$$

To normalize this equation, we choose $\tau = \sqrt{l/g}$ and $\theta_{\max} = \theta_0$ to yield

$$\frac{d^2y}{dx^2} + y = 0, \qquad y(0) = 1.$$

The time and amplitude scales for the problem are now known. Even before we seek a solution, this is important design information if we are attempting to design a pendulum.

This simple linear system has a solution $y = \cos x$ in the transformed space which can be inverse-transformed back to the original problem space to yield the solution

$$\theta = \theta_0 \cos\sqrt{\frac{g}{l}}t.$$

Note that unlike the final solution for $\theta(t)$, the normalized solution for $y(x)$ is independent of the initial conditions or any physical parameters of the pendulum or the gravitational field.

2.6. Normalization of Equations

EXAMPLE 2.6: SPRING-MASS-DASHPOT SYSTEM

Consider the differential for a simple mass supported by a parallel spring and dashpot

$$m\ddot{x} + c\dot{x} + kx = f, \qquad x(0) = 0, \quad \dot{x}(0) = v_0. \tag{2.102}$$

The normalization is begun by introducing the simple scaling transformations $X = x/x_m$, $T = t/\tau$, and $F = f/f_0$. This yields

$$\frac{mx_m}{\tau^2}\ddot{X} + \frac{cx_m}{\tau}\dot{X} + kx_m X = f_0 F, \qquad X(0) = 0, \quad \dot{X}(0) = \frac{v_0 \tau}{x_m}. \tag{2.103}$$

The next step is accomplished by dividing Eq. (2.103) by the coefficient of the linear term, kx_m, to obtain

$$\frac{m}{k\tau^2}\ddot{X} + \frac{c}{k\tau}\dot{X} + X = \frac{f_0}{kx_m}F. \tag{2.104}$$

This step forces the linear term containing the dependent variable to have a unit coefficient and thus to make the magnitude of that term of order one. This is assured because the scaling variables x_m and τ have been chosen so that the new dependent variable, X, and its derivatives, $\dot{X}, \ddot{X}$, will always be of the order of magnitude of one. Because of the manner in which the transformation was defined, every term and every coefficient in Eq. (2.104) is dimensionless. If each term in this differential equation is to contribute in like measure, then each coefficient must be of magnitude one.

At this stage, if the behavior of the system and the parameter values are known, it is possible to compute the coefficients directly and determine the relative importance of each term. If, however, the model is to be used to tailor the response of the system to some desired criteria, as is usual in the process of design, the modeler will desire an analytical approach to normalization and leave the actual performance analysis until the final step.

There are various choices in completing the normalization of even this elementary linear governing equation. These are enumerated below, but all of the approaches are based on the relative importance of the terms in the differential equation. This is an important step, because it involves expert knowledge that the modeler brings to the problem. When this knowledge base has not been established, as when the modeler first approaches a problem or when the analysis is to be used to design some new system, the technique of normalization can still be used to determine the relative importance of the terms in the governing equations.

Small inertia. Spring-mass-dashpot systems having small inertial effects are handled by assuming that the elastic and viscous terms are predominant. Therefore, if the damping term and the forcing-function terms are to be of the same order of magnitude as the linear X term, then the coefficients of these terms must be forced to be of unit value. In this example, this requires assigning values to the time scaling, $\tau = c/k$, and the amplitude scaling, $f_0 = kx_m$. These transform the differential equation and initial conditions to

$$\frac{mk}{c^2}\ddot{X} + \dot{X} + X = F, \qquad X(0) = 0, \quad \dot{X}(0) = \frac{cv_0}{kx_m}. \tag{2.105}$$

Note that this differential equation has a single free parameter, mk/c. The initial conditions, however, also have been affected by the scaling. The initial condition also may be scaled to unity by setting $x_m = cv_0/k$:

$$\frac{mk}{c^2}\ddot{X} + \dot{X} + X = F, \qquad X(0) = 0, \quad \dot{X}(0) = 1. \tag{2.106}$$

If the assumption that the inertial term was negligible was correct, then the coefficient mk/c^2 will be much less than one. If it is much less than one, it may be possible to eliminate this term altogether from the analysis. Note that the normalization permits one to solve a single differential equation, irrespective of the initial-condition value, v_0, or the forcing-function amplitude, f_0. The solution to Eq. (2.106) is therefore a universal solution for this problem although it must be transformed back to the original space if real physical numbers are necessary.

The values of the force, spatial, and temporal scaling factors are, however, of great importance and significance. The analytic expressions for these entities provide insight into the actual time periods over which the majority of changes in the dependent variable occur and the amplitude of these changes. The normalization process therefore yields important information about the problem, even before any attempt is made at obtaining a solution. These expressions are beneficial to the designer wishing to tailor the response of the system and will allow the selection of parameter values that yield acceptable time and amplitude response.

It is not, in general, possible to reduce the governing differential equation to a universal equation having a generic solution for any arbitrary nonlinear problem. The process does, however, reduce the number of free parameters to a minimum, thereby simplifying any parametric analysis.

Small damping. Spring-mass-dashpot systems having small damping effects are handled by assuming that the elastic and inertial terms are predominant. From Eq. (2.104), if the inertial term and forcing functions are to be important, it is necessary to choose their coefficients to be of unit value by setting $\tau^2 = m/k$ and $f_0 = kx_m$. The initial condition is normalized by selecting $x_m = \sqrt{mv_0^2/k}$ which transforms the system to

$$\ddot{X} + c\sqrt{\frac{m}{k}}\dot{X} + X = F, \qquad X(0) = 0, \quad \dot{X}(0) = 1. \tag{2.107}$$

A dimensionless damping factor equivalent to twice the common damping ratio is defined as $\eta = c\sqrt{m/k}$, and the differential equation reduces to

$$\ddot{X} + \eta\dot{X} + X = F, \qquad X(0) = 0, \quad \dot{X}(0) = 1. \tag{2.108}$$

If the remaining coefficient in the differential equation, η, is evaluated using known physical constants and found to be greater than one, then the assumption that the damping was negligible is incorrect and the problem must be reformulated. If it is much less than one, it may be possible to eliminate this term and greatly simplify the analysis. As in the preceeding small-inertia example, the differential equation has a single free parameter and the initial conditions are normalized. Typical *universal* solutions to this problem for a system with a constant forcing function and for various values of η are shown in Fig. 2.18. Note that for all values of η, the initial slopes (velocities) are all equal ($\dot{X}(0) = 1$), and the asymptote of the

2.6. Normalization of Equations

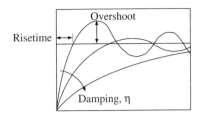

Figure 2.18. Response of a normalized spring-mass-damping system with varying damping.

solutions is $X(\infty) = 1$. The characteristic time over which the major changes occur in X for the smallest value of η is $T = 1$. As η increases to values much greater that one, the characteristic time also must change. In this instance, the assumption of small damping is no longer valid and the problem approaches that of the case of small damping where τ is characterized by c/k, not k/m.

Small spring. A system having a weak elastic element is dominated by inertial and viscous forces. In this instance, the analysis is different than before and the modeler has the flexibility to initially choose the term most likely to dominate the analysis before dividing by its coefficient. If the inertial term is assumed to be a dominant term, then

$$\ddot{X} + \frac{c\tau}{m}\dot{X} + \frac{k\tau^2}{m}X = \frac{f_0\tau^2}{mx_m}F, \qquad X(0) = 0, \quad \dot{X}(0) = \frac{v_0\tau}{x_m}. \tag{2.109}$$

If the viscous term is also dominant, we set its coefficient to unity by choosing $\tau = m/c$. The scaling of the forcing function also can be normalized to yield $f_0 = c^2 x_m/m$ and the initial condition can be normalized by setting $x_m = mv_0/c$.

EXAMPLE 2.7: TEMPERATURE-DEPENDENT RESISTANCE MODEL.

The following is a mathematical model for a resistance-capacitance (RC) circuit having a temperature-dependent resistance element. A schematic of this circuit is shown in Fig. 2.19. In this modeling study, the first attempt will be to model a much simpler system and then extend the model.

Constant Resistance Analysis

Using Kirchoff's laws, one may write the first-order differential equation for the current as

$$R_0 C \frac{di}{dt} + i = C \frac{dE_{in}}{dt}, \qquad i(0) = I_0. \tag{2.110}$$

We next define normalization transformations such that the changes in the dependent and independent variables assume values between 0 and 1. This also assures us that

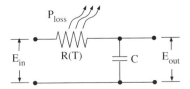

Figure 2.19. Schematic of a temperature-dependent RC network.

derivatives are of order one as well. The transformations assumed here are

$$I = \frac{i}{I_0}, \qquad E = \frac{E_{\text{in}}}{E_0}, \qquad y = \frac{t}{\tau}. \tag{2.111}$$

It is noteworthy that only the time has an unknown scaling factor, τ, because the input voltage is known and its amplitude, E_0, also is given. These variables transform the governing differential equation and initial condition to obtain

$$\frac{R_0 C}{\tau}\frac{dI}{dy} + I = \frac{C}{I_0 \tau}\frac{dE_{\text{in}}}{dy}, \qquad I(0) = 1. \tag{2.112}$$

If each of the terms in this governing differential equation are to be of importance, then we may choose the scaling factor $\tau = R_0 C$, reducing the governing equation to a much simpler form,

$$\frac{dI}{dy} + I = \frac{E_0}{I_0 R_0}\frac{dE}{dy}, \qquad I(0) = 1. \tag{2.113}$$

Since the capacitor acts essentially as a zero resistance if E is changed rapidly, one can solve for the relationship $E_0 = I_0 R_0$ which reduces the equation to

$$\frac{dI}{dy} + I = \frac{dE}{dy}, \qquad I(0) = 1. \tag{2.114}$$

If the input voltage, E_{in}, is given a step increment, E_0, then one may write $dE_{\text{in}}/dt = dE/dy = 0$ and the differential equation has the solution

$$I = e^{-y}. \tag{2.115}$$

The response or normalized current, I, resulting from a unit change in the normalized input voltage, E, is shown in Fig. 2.20. Note that the change of the normalized current, I, is of order one over a change in the normalized time, y, of order one. Thus, the timescale $\tau = R_0 C$, also called the *time constant* of this first-order system, characterizes the temporal changes of this linear first-order system.

The value of this simple model is that we now appreciate the importance of the timescale of the electrical circuit without the complications of a temperature-dependent resistance. We also have gained experience in scaling the problem and finding out the importance of the variables involved and how groups of terms appear in the modeling equations.

Since this is a linear system, the amplitude of the current, I_0, is only a function of the forcing function for the problem, E_0/R_0. Because they appear as individual entities in the modeling equations, one might assume that the resistance, capacitance,

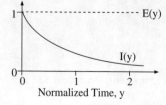

Figure 2.20. Normalized first-order response ($R = $ constant).

2.6. Normalization of Equations

and amplitude of input voltage make this problem a three-degree-of-freedom problem. After normalization, however, it is seen that the problem is only affected globally by one independent parameter, τ, a product of R_0 and C. The amplitude of the response, I_0, is linearly dependent on the ratio E_0/R_0. It is noteworthy that the individual values of R_0 and C are not important to the time response, but that their product directly determines the time over which the majority of change in the current occurs.

Temperature-Dependent Resistance Analysis

The next step in the modeling process is to improve on the initial simple model by including the effect of the temperature change on the resistance element in the circuit. The temperature effect on resistance is modeled using a power-law equation as given by

$$R(T) = R_0 \left(\frac{T}{T_0}\right)^2, \tag{2.116}$$

where the resistance, R, is a function of the absolute temperature, T, and has a resistance of R_0 at some temperature T_0. It now becomes necessary to write an additional equation describing how the temperature of the resistance element varies with time. This is undoubtedly a heat transfer problem, and an equation is thus sought to define the relationship between the electrical energy input, the heat transfer loss, and the temperature of the resistance. This energy balance can be written as

$$P_{\text{in}} - \dot{Q}_{\text{loss}} = mC_p \frac{dT}{dt}. \tag{2.117}$$

Using Kirchoff's laws and neglecting any current flow to the output, it is possible to derive the following governing differential equations describing the current and temperature response to a step change in E_{in}:

$$R_0 C \left(\frac{T}{T_0}\right)^2 \frac{di}{dt} = C \frac{dE_{\text{in}}}{dt} - i\left(\frac{2R_0 C}{T_0} \frac{T}{T_0} \frac{dT}{dt} + 1\right), \tag{2.118}$$

$$\frac{mC_p}{hA} \frac{dT}{dt} = i^2 \frac{R_0}{hA} \left(\frac{T}{T_0}\right)^2 - (T - T_0). \tag{2.119}$$

These equations are transformed using the relationships defined previously and an additional variable to scale the temperature:

$$\theta = \frac{T}{T_0}, \tag{2.120}$$

$$I = \frac{i}{I_0}, \tag{2.121}$$

$$y = \frac{t}{\tau}, \tag{2.122}$$

$$E = \frac{E_{\text{in}}}{E_0}. \tag{2.123}$$

These transformations yield the normalized governing nonlinear first-order differential equations

$$\frac{R_0 C}{\tau} \theta^2 \frac{dI}{dy} = \frac{E_0}{I_0 R_0} \frac{R_0 C}{\tau} \frac{dE}{dy} - I\left[2\frac{R_0 C}{\tau} \theta \frac{d\theta}{dy} + 1\right], \tag{2.124}$$

$$\frac{(mC_p/hA)}{\tau}\frac{d\theta}{dy} = I^2 \frac{I_0^2 R_0}{hAT_0}\theta^2 - (\theta - 1). \tag{2.125}$$

The selection of the values of the scaling variable τ is based on the concept that each important term in the normalized governing differential equations is of order one. Thus, if both the derivatives for θ and I are of similar magnitude, their coefficients must be approximately of the same order. We are still left with the problem of deciding on the value of τ as either $R_0 C$ or mC_p/hA. The choice of either should be based on the notion that it characterizes the time during which the greatest changes occur. It thus is assumed for this example that the time during which the current change occurs is smaller than that for the temperature change. We therefore choose this timescale to normalize the temporal variations

$$\tau = R_0 C. \tag{2.126}$$

We may also select a value for $E_0 = I_0 R_0$ and the resultant normalized differential equations are thus

$$\theta^2 \frac{dI}{dy} = \frac{dE}{dy} - I\left[2\theta \frac{d\theta}{dy} + 1\right], \tag{2.127}$$

$$\frac{d\theta}{dy} = I^2 \left(\frac{I_0^2 R_0}{hAT_0}\right)\left(\frac{hA\tau}{mC_p}\right)\theta^2 - \left(\frac{hA\tau}{mC_p}\right)(\theta - 1). \tag{2.128}$$

The importance of the coefficients of the terms in the second differential equation lies in the relative magnitude of the ratios

$$R_\tau = \left(\frac{\tau}{(mC_p/hA)}\right), \tag{2.129}$$

$$R_{\text{pwr}} = \left(\frac{I_0^2 R_0}{mC_p T_0}\right). \tag{2.130}$$

Introducing these into the governing equations results in

$$\theta^2 \frac{dI}{dy} = \frac{dE}{dy} - I\left[2\theta \frac{d\theta}{dy} + 1\right], \tag{2.131}$$

$$\frac{d\theta}{dy} = R_{\text{pwr}} I^2 \theta^2 - R_\tau (\theta - 1). \tag{2.132}$$

The first of these ratios, R_τ, represents the relative timescales of the electrical and thermal problems. If they are of the same magnitude, then this ratio is of order one and the terms that this ratio prefixes are important to the problem. If the thermal inertia is such that the current changes much faster than the temperature, then these same terms may be neglected. In this case, $R_\tau \ll 1$, inferring that the temperature is essentially constant since the right side of the $d\theta/dy$ equation is multiplied by this small term.

The second of the ratios, R_{pwr}, represents the relative ability of the system to dissipate power through $i^2 R$ losses and the ability to absorb this power as thermal energy.

The solution of these two equations is not straightforward. Large nonlinearities appear as the terms θI^3, $I\theta^{-3}$, and $I^2\theta^2$. A solution to this system of equations may be obtained for specific data for the physical problem. In this example, the data shown in Table 2.1

2.6. Normalization of Equations

TABLE 2.1. Property Data for the Resistor and Its Heat Transfer Coefficient

Variable	Value	Units
Heat transfer coefficient, h	10.0	W/s-cm^2-°C
Surface area, A	3.0	cm^2
Mass of resistor, m	10.0	g
Specific heat, C_p	0.3	W/g-°C
Nominal temperature, T_0	23.0	°C
Nominal resistance, R_0	100.0	kΩ
Capacitance, C	0.1	μF (microfarads)
Input voltage, E_i	10.0	V
Time constant ratio, R_τ	0.001	s
Power ratio, R_{pwr}	0.1	

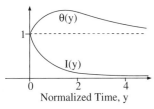

Figure 2.21. Coupled normalized current and temperature response of a temperature-dependent resistance.

will be used to obtain a numerical solution. The solution for the transient response of the system variables is portrayed in normalized form in Fig. 2.21. From this result, it can be observed that the current changes are of the order of one but that the temperature changes are much smaller than one. This infers that the current has been properly scaled [Eq. (2.121)] but that the temperature has not [Eq. (1.120)]. This variable remains very close to unity throughout and its derivatives are obviously much smaller than unity. In addition, it is noted that the dimensionless timescales for the current and temperature changes are of the order of 1 and 20, respectively. The previous decision to scale the problem on the current was therefore an appropriate one.

Additional methods of approach may be used to obtain solutions to this problem. One may use the solution information to reduce the complexity of the problem by setting $\theta = 1$ which decouples the current equation and permits direct analytic solution for $I(y)$. An improvement in the method of normalization used would be to seek a more appropriate scaling for the temperature and rework the numerical solution to obtain higher accuracy for both solution variables, $I(y)$ and $\theta(y)$. As an example, a better choice for the scaling of θ might have been a difference term such as

$$\theta' = \frac{T - T_0}{T_m - T_0}. \tag{2.133}$$

This has the advantage of introducing a new constant, the maximum expected temperature, T_m, into the governing equations. A similar approach should be used to scale the current flow. The student is invited to explore these ideas to gain insight into the normalization technique and into the physics of this problem.

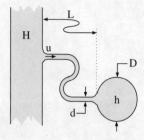

Figure 2.22. Diagram of the balloon-catheter system to be analyzed.

EXAMPLE 2.8: CATHETER-BALLOON SYSTEM MODEL

Consider a small inflatable balloon on the end of a rigid tube. The physical situation is depicted in Fig. 2.22. There are many such physical devices used in medicine. In these devices, the balloon usually is made of latex and is inflatable. The tube is usually polyethylene and is relatively nondistensible. The governing differential equations to be used to simulate the response of this system are

$$L\frac{du}{dt} = g(H - h) - \frac{fL}{2d}u\,u\,, \tag{2.134}$$

$$\frac{\pi d^2}{4}u = \frac{d(\text{Vol})}{dt} = \frac{\pi}{2}D^2\frac{dD}{dt}, \tag{2.135}$$

$$h = 2C\frac{w}{D_0}\left[\frac{D}{D_0} - 1\right]\left[\frac{D_0}{D}\right]^2, \tag{2.136}$$

where u = velocity in tube, f = friction factor, L = length of tube, H = inlet pressure, h = balloon pressure, g = local acceleration of gravity, d = catheter tube diameter, D = balloon diameter, D_0 = initial balloon diameter, w = balloon wall thickness, and C = material constant. It is now possible to analyze the system and evaluate its response. The following steps are to be undertaken:

1. Nondimensionalize the time with respect to the scaling variable t_f, the balloon diameter and velocity w.r.t. their initial values, D_0 and U_0. These definitions are to be inserted into the governing equations, transforming them into more usable form.
2. Use numerical solution techniques to evaluate the response of the balloon to a step change of the pressure, h.

The first task is to define nondimensional variables, and time is first represented as

$$T = \frac{t - t_0}{t_f}, \tag{2.137}$$

where t_0 and t_f are the starting and finishing times, respectively, of our analysis. The dependent variables may be represented similarly. The velocity is arbitrarily scaled in a similar manner using the boundary conditions that at $t = 0$, $h = h_0$, $u = U_0$, and the initial condition for velocity is $du/dt = 0$:

$$L\frac{du}{dt}\bigg|_{t=0} = g(H - h_0) - \frac{f}{2}\frac{L}{d}U_0|U_0| = 0 \tag{2.138}$$

2.6. Normalization of Equations

or

$$U_0 = \sqrt{\frac{2dg}{fL}(H - h_0)}. \tag{2.139}$$

The dimensionless velocity, V, and balloon diameter, D, are simply scaled as fractions of their initial values. The pressure in the balloon, h, is scaled as a fraction of the upstream pressure head, H. These are defined as

$$V = u/U_0, \tag{2.140}$$
$$D = D/D_0, \tag{2.141}$$
$$P = h/H. \tag{2.142}$$

The differential equation for $u(t)$ is therefore transformed to an equation in terms of $V(T)$. At $t_0 = 0$, this becomes

$$\frac{dV}{dT} = \frac{gHt_f}{U_0 L}(1 - P) - \frac{fU_0 t_f}{2d} V|V|. \tag{2.143}$$

If the rate of change of volume of the balloon is equated to the incoming flow rate (continuity equation), an equation for the rate of change of the balloon diameter, D, is obtained:

$$D^2 \frac{dD}{dt} = \frac{U_0 d t_f}{2 D_0^2} V, \tag{2.144}$$

where the balloon pressure is given by

$$P = \frac{2Cw}{HD_0}\left[\frac{D-1}{D^2}\right]. \tag{2.145}$$

Since t_f has not been defined, this is now arbitrarily chosen to make the terms in one of the differential equations unity. Thus,

$$t_f = \frac{U_0 L}{gH}, \tag{2.146}$$

which reduces Eq. (2.143) to a much simpler form:

$$\frac{dV}{dT} = (1 - P) - f\left(\frac{U_0^2}{2gH}\right)\left(\frac{L}{d}\right) V|V|. \tag{2.147}$$

The differential equation for the balloon diameter, Eq. (2.144), is similarly reduced to

$$D^2 \frac{dD}{dT} = \left[\frac{U_0^2}{2gH}\right]\left[\frac{d}{D_0}\right]^2 \left[\frac{L}{d}\right] V. \tag{2.148}$$

Note that all of the above terms in Eqs. (2.147) and (2.148) are dimensionless. The problem is now normalized, and the response of the system may be determined readily using any one of a number of schemes. It is even possible to achieve a solution to a forced system where the balloon pressure might be time varying. In such a situation, the means by which one imposes this pressure should be carefully worked into the solution technique. In addition, the forcing function also must be normalized.

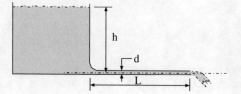

Figure 2.23. Pipe draining a reservoir.

An added feature of this method is that the relative contributions of the individual terms in the differential equation are now readily apparent. In addition, if one employs numerical solution techniques, the magnitude of the step size to be employed is now bounded and has the property $\delta \ll 1$.

EXAMPLE 2.9: PIPE DRAINING A RESERVOIR

Consider the long pipe draining a large reservoir as shown in Fig. 2.23. The initial depth of the water is h, and the pipes have a diameter d and length L. What is sought here is the flow velocity at the open end of the drain pipe as a function of time, $v_e(t)$. As a first step in modeling this system, we make the following assumptions:

1. incompressible, frictionless flow;
2. flow from the open surface to the exit is considered as streamline flow;
3. negligible changes in atmospheric pressure with location or altitude, z;
4. $v_{\text{surf}} \approx 0$; and
5. the tank height, h, is approximately constant.

Under these assumptions, it can be shown that

$$\frac{v_e^2}{2} + L\frac{dv_e}{dt} = gh, \qquad v_e(0) = 0. \tag{2.149}$$

We next introduce the scaling variables, $V = v_e/v_{\text{max}}$ and $T = t/\tau$, and the differential equation becomes

$$\frac{2L}{\tau v_{\text{max}}}\frac{dV}{dt} + V^2 = \frac{2gh}{v_{\text{max}}^2}, \qquad V(0) = 0. \tag{2.150}$$

If we choose $v_{\text{max}} = \sqrt{2gh}$, $\tau = 2L/v_{\text{max}}$, the equation is further reduced to an extremely simple form,

$$\frac{dV}{dt} + V^2 = 1, \qquad V(0) = 0. \tag{2.151}$$

The solution to the transformed governing equation is thus

$$V = \tanh(T) \tag{2.152}$$

and appears graphically in Fig. 2.24. As can be seen, there is only a single solution for the transformed problem. In this domain, the problem is unique and independent of **any** of the problem's parameter values. This is directly the result of scaling the domain. As a

2.6. Normalization of Equations

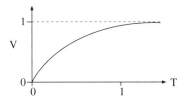

Figure 2.24. Normalized reservoir-pipe exit velocity versus time.

general rule, the scaling transformations are functions of the parameters of the system. Thus, in this example, as the tank height, pipe length, or local acceleration of gravity change, the spatial and temporal domains are stretched or compressed according to the definitions of $v_{\max}$ and τ, respectively.

2.6.1 Complex Nonlinear Problems

Frequently, one encounters complex nonlinear systems such as

$$\frac{dv}{dt}(a+bv) = c\sqrt{\frac{v-v_0}{v}}. \tag{2.153}$$

This can be transformed to the form

$$\frac{cv_0}{\tau}\frac{dy}{dx}(d+ey) = \sqrt{\frac{y-1}{y}}. \tag{2.154}$$

In this problem, the most probable choice of τ would be

$$\tau = cev_0. \tag{2.155}$$

In this choice, the coefficient of the nonlinear product term of $y\,dy/dx$ would be set to unity. This is done because both dy/dx and y are assumed to be of importance in this problem. The resulting differential equation is thus

$$\frac{dy}{dx}(g+y) = \sqrt{\frac{y-1}{y}}. \tag{2.156}$$

At this juncture, one must compute the size of the constant term g. If it has a magnitude much greater than one, then the choice of the time constant must be revised to $\tau = cdv_0$, yielding the equation

$$\frac{dy}{dx}(1+hy) = \sqrt{\frac{y-1}{y}}, \tag{2.157}$$

where the constant h is now much less than one.

2.6.2 Infinite and Semi-Infinite Domains

The simple linear transformation method described in the beginning of this section cannot be applied when one or more of the domains is infinite or semi-infinite in extent. In these cases, there is no meaning to a characteristic length, and a more novel approach must be taken that will transform the domain to a finite region. This is, perhaps, best understood through the following example.

EXAMPLE 2.10: SEMI-INFINITE DOMAIN TRANSFORMATION

Consider the differential equation,

$$(1+y)\frac{dy}{dt} = a, \qquad y(0) = 0. \tag{2.158}$$

In this system, the derivative vanishes slowly as $t \to \infty$, and y therefore grows without bound. If one were to attempt the usual transformation variables, the problem would not be amenable to the change in variables; it is impossible to discern a spatial characteristic value for the problem because of the infinite domain in y. To transform this problem successfully, however, we introduce a new set of nonlinear transformations as given by

$$\eta = \frac{1}{y+1}, \qquad T = t/\tau. \tag{2.159}$$

Here, the new variable η is chosen to remain bounded within $0 \le \eta \le 1$ as $y \to \infty$. The differential equation following transformation then becomes

$$\frac{d\eta}{dT} + a\tau\eta^3 = 0, \qquad \eta(0) = 1. \tag{2.160}$$

By choosing $\tau = 1/a$, this equation may be separated and integrated to yield the transformed and untransformed solutions,

$$T = \frac{1}{2}\left(\frac{1}{\eta^2} - 1\right), \qquad y = \sqrt{2at+1} - 1. \tag{2.161}$$

Here, the resultant steady state is at $\eta = 0$, and the time constant takes on a meaning as the length of time required for y to undergo its greatest changes in the original problem domain.

2.7 TRANSFORMATION OF VARIABLES

In many problems, there are sound mathematical reasons to change the formulation of the variables of the problem. In Section 2.6, this was done using a set of simple transformations that resulted in a magnification or reduction of the size and an offset from the original variable. This is typified in Eq. (2.162), where the variable x is offset by some value x_0 and linearly scaled by the expected variation in x given by Δx:

$$\eta = \frac{x - x_0}{\Delta x}. \tag{2.162}$$

By introducing such linear transformations, all variables can be scaled to be of the order of one (normalization transformation).

2.7. Transformation of Variables

The transformation of Eq. (2.162) is a simple linear example of a more general approach. The concept of a general transformation of variables that can be tailored to the specific needs of the modeler is introduced next. If only the independent variables were transformed, the domain of the problem would become warped, resulting in an alteration of the functional relationship between the dependent variable and its new domain. Thus, some function $U(x, y, z, t)$ might be transformed to $U(\alpha, \beta, \gamma, \delta)$. It is not essential that the dimension of the domain remain fixed. It may be possible to define transformations such that

$$u(x, y, z, t) \Rightarrow U(\eta, \zeta), \tag{2.163}$$

where the new variables are functionally mapped to the new domain. An example of this form of transformation would be the change of variable from spherical to planar coordinates as

$$z(r, \theta, \phi) \Rightarrow z(x, y). \tag{2.164}$$

This is a common mapping transformation used to portray the spherical surface of Earth as a flat plane. It is obvious that this transformation is not uniform. On most maps, the regions near the poles of Earth are stretched, whereas the areas near the equator are unchanged. In fact, if the transformation is extended all the way to the poles, each point representing one of the two poles becomes a line of finite length. Section 2.7.1 introduces the mathematics associated with such transformations.

2.7.1 Exponential Transformations

Differential equations having nonlinearities of the form

$$\frac{dx}{dt} = f(x^n, t) \tag{2.165}$$

often can be transformed into a linear domain, $z(T)$, using the transformation

$$x = x_0 z^p, \qquad T = t/\tau. \tag{2.166}$$

The differential equation then becomes

$$\frac{dx}{dt} = \frac{p x_0 z^{p-1}}{\tau} \frac{dz}{dT} = f(z^{pn}, T). \tag{2.167}$$

In many cases, this can be reduced further by the proper choice of the exponent, p, to yield a linear differential equation of the form

$$\frac{dx}{dT} = f(z, T). \tag{2.168}$$

This transformation will not work if the function, $f(x^n, t)$, is of the form $f(x^n, t) = c + g(x^n, t)$, where c is a constant.

EXAMPLE 2.11: NONLINEAR SPRING-DASHPOT

Consider the differential equation for this system given as

$$c\frac{dx}{dt} + kx^2 = 0, \qquad x(0) = x_0. \tag{2.169}$$

We define $x = x_0 z^p$ and $T = t/\tau$ to transform the domain to

$$\frac{dx}{dt} = \frac{x_0 p z^{p-1}}{\tau}\frac{dz}{dT} = -\frac{k}{c}x^2 = -\frac{kx_0^2}{c}z^{2p}, \tag{2.170}$$

which can be simplified to obtain

$$\frac{dz}{dT} = -\frac{k\tau}{pc}z^{p+1}, \qquad z(0) = 1^{1/p}. \tag{2.171}$$

By choosing $p = -1$ and $\tau = c/(kx_0)$, the differential equation is reduced to

$$\frac{dz}{dT} = 1, \qquad z(0) = 1. \tag{2.172}$$

This is readily solved to obtain

$$z = z(0) + T = 1 + T, \tag{2.173}$$

or, returning to the original domain, the exact solution can be shown to be

$$\frac{x}{x_0} = \frac{c}{c + kx_0 t}. \tag{2.174}$$

2.7.2 Generalized Transformation

The transformation of any variable or even a system of equations can be readily demonstrated mathematically. Equation (2.162) is one such transformation. Generally, however, the transformation of a dependent variable, $u(x_i)$, in an I-dimensional domain to a new variable, $v(y_j)$, within a J-dimensional domain is more than mere substitution. Often, the equations to be transformed are integral or differential. We begin by explicitly specifying the relationships defining the transformation:

$$u = u(v),$$
$$y_1 = y_1(x_i),$$
$$y_2 = y_2(x_i),$$
$$\vdots$$
$$y_J = y_n(x_i),$$

where $x_i = [x_1, x_2, \ldots, x_I]$. The chain rule allows us to build the derivatives essential to transforming differential equations. Thus,

$$\frac{\partial u}{\partial x_1} = \frac{\partial u}{\partial v}\left[\frac{\partial v}{\partial y_1}\frac{\partial y_1}{\partial x_1} + \frac{\partial v}{\partial y_2}\frac{\partial y_2}{\partial x_1} + \cdots + \frac{\partial v}{\partial y_J}\frac{\partial y_J}{\partial x_1}\right]. \tag{2.175}$$

The transformations introduced in Section 2.6 are linear examples of this class of transformations.

2.7.3 Similarity Transformations

Many solutions to problems in the general area of mechanics can be obtained by introducing a class of transformations using similarity variables. As an example, in some vibrations problems, the amplitude of vibration at a given location may be identical to that at other locations but scaled by a spatially dependent factor. In the problem of laminar flow over a flat plate, the x components of the velocity, u, at two different locations are merely scaled versions of each other. The common factor between these two situations is that their mathematical solution may be derived from a transformation of the variables. The mathematical formalism for this class of transformations was introduced by Abbott [1].

In this class of transformations, a new domain for the problem is defined that will reduce the number of independent variables in the differential equation. One of the new variables is termed a separation variable because it permits the method of separation of variables to be applied to the resulting differential equation. Such a variable, if it can be found, must not only permit separation of variables to be applied, it must also satisfy the initial and boundary conditions for the physical problem.

The problems that may be solved using this method may be classified into two broad categories. The first category includes problems that are well posed. For well-posed problems, the differential equations and boundary conditions are sufficient to ensure that a unique solution exists and the solution must be continuously dependent on the boundary conditions. Those problems for which these conditions are not met are ill-posed problems. There is no formal method available by which to derive the form of these transformations. One may only assume a transformation formulation and test its applicability and the uniqueness of the solution.

For the well-posed problem, it is possible to state [1] that only one solution will result for an assumed transformation, if a solution can be found. The transformation might not be one that would guarantee similarity and transform the problem to a domain where a solution is possible. If the problem were not sufficiently well defined by the omission of some boundary or initial conditions, then even if the assumed transformation were appropriately chosen and the transformation was to a domain where a similarity solution exists, there is still no guarantee of uniqueness of the solution, if it could be found. As an example, consider the simple differential equation

$$\frac{dx}{dt} = 5. \tag{2.176}$$

Unless an initial condition is specified, an infinite number of solutions exists.

There cannot be more than a single independent similarity variable for a well-posed problem for any given class of transformation because only one solution exists and any other similarity variable must be related uniquely to the first by the solution.

Throughout the remainder of this chapter, there is no attempt to distinguish between the physical boundaries of a problem and its temporal initial conditions. Only the term "boundary conditions" is applied henceforth, but it must be remembered that this is inclusive of any initial conditions for the problem.

The method introduced in the following sections may reduce a problem from a partial differential equation in two independent variables to an ordinary differential equation in a single variable. Obviously, the order of the equation will increase to allow one to apply the boundary conditions of the original problem. It may be possible to iteratively apply this method to higher-order equations to reduce them to ordinary differential equations.

Consider the problem of a suddenly accelerated flat plate moving in its own plane (the xz plane). The infinite plate is instantly brought from rest to a constant velocity, U_0, in the x direction. The surrounding fluid is stationary with respect to the coordinate system. This problem can be solved exactly for the two-dimensional case. In this case, the only nonzero component of the velocity is that parallel to the plate (x direction). The Navier-Stokes equations for an incompressible fluid under these conditions reduce to

$$\frac{\partial u}{\partial t} = \nu \frac{\partial^2 u}{\partial y^2}, \tag{2.177}$$

subject to the boundary conditions

$$\begin{aligned} t &= 0 \quad \text{and} \quad y \geq 0 : u = 0, \\ t &= 0 \quad \text{and} \quad y > 0 : u = U_0, \\ t &= \infty \quad \text{and} \quad y \geq 0 : u = 0. \end{aligned} \tag{2.178}$$

Assuming that the solution is separable, that is,

$$u = T(t)Y(y), \tag{2.179}$$

the governing equation becomes (for T, Y not identically zero)

$$\frac{T'}{T} = \nu \frac{Y''}{Y} = \text{constant} = -\lambda. \tag{2.180}$$

Defining $\omega = \sqrt{\lambda/\nu}$, the two resulting solutions are then

$$T = T_0 e^{-\lambda t}, \tag{2.181}$$

$$Y = Y_0 \cos \sqrt{\omega} y + Y_1 \frac{\sin \omega y}{\omega}, \tag{2.182}$$

resulting in the velocity solution

$$u = e^{-\lambda t} \left[U_1 \cos \omega y + \frac{U_2 \sin \omega y}{\omega} \right]. \tag{2.183}$$

When the boundary conditions of Eq. (2.178) are applied, the solution is $u = $ constant. Therefore, the classical method of separation of variables only provides a trivial solution

2.7. Transformation of Variables

and some other means must be found to obtain a solution. By shifting the coordinate system, however, it is possible to demonstrate that the method of separation of variables will work.

In the two solutions presented in the following sections, the first assumes that the problem is well posed. The second assumes that the initial conditions are not known.

2.7.4 Well-Posed Problems

This method assumes that a functional relationship or transformation exists between the dependent variable in the x, y space, $u(y, t)$, and the same variable in some new space. Two new independent variables are defined but one of these is a primitive variable with a trivial definition, $\xi = t$. The result is that the remaining variable, η, contains all of the dimensional information of the original variables. This is equivalent to having a single independent variable, η, producing a mapping of the form

$$u(x, y) \longrightarrow u(\eta). \tag{2.184}$$

This is accomplished using the transformation $\eta = \eta(y, x)$. This *de facto* reduction in dimension is necessarily accompanied by a reduction in the number of boundary conditions. The transformation must interrelate the original boundary conditions so that they are satisfied. The condition is equivalent to

$$u(\eta[c, y]) = u(\eta[x, d]). \tag{2.185}$$

If this condition cannot be met, then no similarity transformation exists. In boundary-layer flow over a flat plate, the boundary conditions reduce to

$$u(x, \infty) = u(0, y) = 0. \tag{2.186}$$

A transformation consistent with this requirement is

$$\eta = (ay^n/x^m). \tag{2.187}$$

The only requirement for this function is that n, m must be real. They may be greater or less than zero. To keep the transformation general for now, the mapping is $(x, y) \rightarrow (\eta, \xi)$. The choice of $\xi = x$ prevents any misinterpretation of the independent variables in the old and new domains. It is also the simplest choice for mapping this variable, but there are no theoretical bases for making a choice of the variable mapping for η. The value of the selection can be ascertained only by carrying through the analysis and observing whether the transformation works. If it fails to yield results, then it can only be stated that the assumed transformation does not yield a similarity solution. Applying the chain rule from calculus,

$$\frac{\partial u}{\partial y} = \frac{\partial u}{\partial \xi}\frac{\partial \xi}{\partial y} + \frac{\partial u}{\partial \eta}\frac{\partial \eta}{\partial y} = an\frac{y^{n-1}}{x^m}\frac{\partial u}{\partial \eta}. \tag{2.188}$$

The choice of n is arbitrary, but it directly affects the form of the resulting solution. The appropriate choice therefore is predicated on the form of the differential equation that results. There is no loss of generality in assuming any particular value of n, and an examination of Eq. (2.188) shows that the value of $n = 1$ greatly reduces the complexity of the problem. Applying the same technique to the other partial differentials in Eq. (2.177) yields

$$\frac{\partial u}{\partial y} = \frac{a}{\xi^m} \frac{\partial u}{\partial \eta},$$

$$\frac{\partial^2 u}{\partial y^2} = \frac{a^2}{\xi^{2m}} \frac{\partial^2 u}{\partial \eta^2},$$

$$\frac{\partial u}{\partial x} = \frac{\partial u}{\partial \xi} - m\frac{\eta}{\xi} \frac{\partial u}{\partial \eta}.$$

These transform the differential equation to

$$\frac{\partial u}{\partial \xi} - m\frac{\eta}{\xi} \frac{\partial u}{\partial \eta} = \nu \frac{a^2}{\xi^{2m}} \frac{\partial^2 u}{\partial \eta^2}. \tag{2.189}$$

The method of separation of variables, assuming $u = bg(\xi) f(\eta)$, yields

$$\xi^{2m} \frac{g'}{g} = m\frac{\eta}{\xi^{1-2m}} \frac{f'}{f} + \nu a^2 \frac{f''}{f}, \tag{2.190}$$

which is separable for $m = 1/2$. The full mapping therefore is based on the transformations

$$\eta = ay/\sqrt{x}, \tag{2.191}$$
$$\xi = x, \tag{2.192}$$

yielding

$$\xi \frac{g'}{g} = \frac{\eta}{2} \frac{f'}{f} + \nu a^2 \frac{f''}{f} = \lambda, \tag{2.193}$$

producing the solution $g(\xi) = c_1 \xi^\lambda$. Applying the boundary condition $u(0, x) = u_0$ produces the conclusion that $\lambda = 0$ and $g = c_1$ (constant). The differential equation now can be written as

$$2\nu a^2 f'' + \eta f' = 0. \tag{2.194}$$

The quantities a, b, and c_1 are all arbitrary constants. If one chooses $a^2 = \nu/4$ and $c_1 b = u_0$, then the differential becomes parameter-free (normalized), or

$$f'' + 2\eta f' = 0, \tag{2.195}$$

where f is only a function of $\eta = \frac{y}{2}\sqrt{\frac{\nu}{x}}$ and the boundary conditions are

$$\eta = 0: \quad f(0) = 1, \tag{2.196}$$
$$\eta = \infty: \quad f(\infty) = 0. \tag{2.197}$$

This governing equation has an exact solution,

$$u = u_0 f(\eta) = u_0 (1 - \text{erf}(\eta)), \tag{2.198}$$

2.7. Transformation of Variables

where erf(η) is the error function given as

$$\text{erf}(\eta) = \frac{2}{\sqrt{\pi}} \int_0^{\eta} e^{-x^2} dx. \tag{2.199}$$

For additional details on this solution, refer to Schlichting [41].

2.7.5 Problems with Missing Boundary Conditions

As a second application of this similarity mapping technique, the boundary-layer problem is repeated, but assuming that the initial condition is not known. The formulation of the problem is thus

$$\frac{\partial u}{\partial x} = \nu \frac{\partial^2 u}{\partial y^2}, \tag{2.200}$$

$$\left. \begin{array}{l} y = 0, \ x > 0 : u = u_0 \\ y \to \infty, \ x > 0 : u = 0 \end{array} \right\}. \tag{2.201}$$

Since all of the boundary conditions are not defined, the condition of similarity given in Eq. (2.185) cannot be used to check the similarity transformation. The general transformation assumed is of the form

$$\xi = x, \tag{2.202}$$

$$\eta = \frac{ay}{\gamma(t)}, \tag{2.203}$$

yielding the differential equation

$$\frac{\partial u}{\partial \xi} - \eta \frac{\gamma'}{\gamma} \frac{\partial u}{\partial \eta} = \frac{\nu a^2}{\gamma^2} \frac{\partial^2 u}{\partial \eta^2}. \tag{2.204}$$

This equation may be changed by applying the method of separation of variables, assuming $u = b f(\eta) g(\xi)$, or

$$\gamma^2 \frac{g'}{g} = \gamma \gamma' \eta \frac{f'}{f} + \nu a^2 \frac{f''}{f} = \lambda, \tag{2.205}$$

which is separable, providing $\gamma \gamma' = c_0$, where c_0 is a constant. The solutions to the differential equations for γ, g are substituted into the expression for the velocity u to obtain

$$u = b c_1 (\xi + \xi_0)^{\lambda/2c_0} f(\eta). \tag{2.206}$$

Now the boundary condition can be applied that $u(y = \eta = 0) = u_0$,

$$u_0 = b c_1 (\xi + \xi_0)^{\lambda/2c_0} f(0), \tag{2.207}$$

which can be satisfied only when $\lambda = 0$ and $g = c_1 =$ constant (same result obtained previously). If $a^2 = \nu/4$, $c - 0 = 1/2$, and $c_1 = u_0$, the similarity transformation variable becomes

$$\eta = \frac{2}{2\sqrt{\nu(x + x_0)}}, \tag{2.208}$$

and the differential equation for the function $f(\eta)$ controlling the velocity $u = u_0 f(\eta)$ reduces to

$$f'' + 2\eta f' = 0, \tag{2.209}$$

subject to the conditions

$$\eta = 0: \quad f(0) = 1, \tag{2.210}$$
$$\eta = \infty: \quad f(\infty) = 0. \tag{2.211}$$

At this point, the solution would follow as in Section 2.7.4. Interestingly, this method infers that it is sometimes possible to get similarity solutions even when all of the boundary conditions are not known. The solution must, of course, agree with the physical problem.

2.7.6 Summary

The two simple analytical techniques presented for generating similarity solutions are applicable to a variety of problems in the area of mechanics. The second method contains a fundamentally different and larger class of transformations than is provided in the first method. It is also noteworthy that this technique may produce one, many, or no similarity variables that will reduce a partial differential equation to an ordinary differential equation. Additionally, the similarity transformation may not meet the boundary conditions. Abbott [1] concludes the following:

1. The existence of a similarity variable implies the lack of a length that can be used to scale one or more dimensions of the problem. This length is termed a *characteristic length*.
2. In some cases it may be possible to determine unique similarity variables by considering only the partial differential equation, even in the absence of all of the boundary conditions. In other cases it is necessary to find a secondary condition arising from known boundary conditions or from conservation theorems.
3. If a similarity variable is not found for the particular class of transformations assumed, it cannot be concluded that none exist. One can only conclude that under the assumed transformation variable, no similarity solution exists.
4. Similarity solutions are often limiting, or asymptotic, solutions to a given problem. This is indicated by the constant x_0 in Eq. (2.208).
5. The methods presented are mathematically a reduction by one in the number of independent variables. It is possible to apply the method repeatedly to obtain a solution to the differential equations.

2.8 UNDERSTANDING DIFFERENTIAL EQUATIONS

It is often possible to understand the type of solution one might get for a nonlinear system by finding a comparable differential equation to a well-known, simpler linear system. The following examples of combinations of a spring element, a damping element, and an inertial element will aid in formulating a general understanding of the terms in a differential equation.

2.8. Understanding Differential Equations

2.8.1 Force on a Spring

Consider the simple equation describing a force acting directly on a spring:

$$kx = F. \qquad (2.212)$$

In this simple linear system, there is a direct and linear relationship between the applied force, F, and the resulting dependent variable, the displacement, x. The application of a steady-state force yields a steady-state solution:

$$x = F/k. \qquad (2.213)$$

For a nonlinear equation, the nonlinear function $f(x)$ for the elastic response of the spring element results in the governing equation,

$$f(x) = F. \qquad (2.214)$$

As before, there is a direct, but in this case a nonlinear, relationship between the applied force and the resulting displacement.

2.8.2 Force on a Damper

For a simple damper, dashpot, or shock absorber with an applied force, the describing equation is a first-order, ordinary differential equation:

$$c\dot{x} = F, \quad x(0) = x_0. \qquad (2.215)$$

For this system, the velocity and the force are linearly related. The steady-state solution only exists where $F \equiv 0$.

2.8.3 Force on a Mass

Newton's law for the response of a mass to an applied force is

$$m\ddot{X} = F. \qquad (2.216)$$

In general, there is no steady-state solution to this problem, even for a zero force.

2.8.4 Force on a Spring-Damper Pair

The combination of these two elements results in the first-order, ordinary linear differential equation

$$c\dot{x} + kx = F. \qquad (2.217)$$

It is seen readily that applying the normalization transformations

$$\begin{aligned} X &= x/x_0, \\ T &= t/\tau, \\ \mathcal{F} &= F/F_0, \end{aligned} \qquad (2.218)$$

results in the following governing equation:

$$\frac{c}{k\tau}\dot{X} + X = \frac{F_0}{kx_0}\mathcal{F}. \tag{2.219}$$

The characteristic time constant for this system is thus $\tau = c/k$ and the characteristic amplitude is $x_0 = F_0/k$. This yields the normalized differential equation

$$\dot{X} + X = \mathcal{F}.$$

2.8.5 Force on a Spring-Mass Pair

The combination of spring and mass elements results in the second-order, linear ordinary differential equation

$$m\ddot{X} + kx = F. \tag{2.220}$$

Using the same transformations as in Eq. (2.218) yields

$$\frac{m}{k\tau^2}\ddot{X} + X = \frac{F_0}{kx_0}\mathcal{F}. \tag{2.221}$$

The characteristic time constant for this system is thus $\tau = \sqrt{(m/k)}$ and the characteristic amplitude is $x_0 = F_0/k$. This system has no damping and therefore has a steady state that is oscillatory. The resulting normalized governing equation is thus

$$\ddot{X} + X = \mathcal{F}.$$

2.8.6 Force on a Mass-Dashpot Pair

The combination of dashpot and mass elements results in the second-order, linear ordinary differential equation

$$m\ddot{X} + c\dot{x} = F. \tag{2.222}$$

Using the same transformations as in Eq. (2.218) yields

$$\frac{m}{c\tau}\ddot{X} + \dot{X} = \frac{F_0\tau}{cx_0}\mathcal{F}. \tag{2.223}$$

The characteristic time constant for this system is thus $\tau = m/c$ and the characteristic amplitude appears as $x_0 = F_0 m/c^2$. Although the system has damping, it does not have an elastic term and therefore lacks a well-defined steady-state response. This typically will be a constant velocity, including the special case of $\dot{x} = v = 0$ when $F = 0$. The normalized governing equation is thus

$$\ddot{X} + \dot{X} = \mathcal{F}.$$

2.8.7 Force on a Spring-Mass-Dashpot System

The combination of spring, dashpot, and mass elements results in the second-order, linear ordinary differential equation

$$m\ddot{x} + c\dot{x} + kx = F. \tag{2.224}$$

Application of the same transformations as in Eq. (2.218) yields

$$\frac{m}{k\tau}\ddot{X} + \frac{c}{\sqrt{mk}}\dot{X} + X = \frac{F_0 \tau}{kx_0}\mathcal{F}. \tag{2.225}$$

The characteristic time constant for this system is thus $\tau = \sqrt{(m/k)}$ and a characteristic amplitude of $x_0 = F_0/k$. Selection of these scaling factors yields

$$\ddot{X} + \xi \dot{X} + X = \mathcal{F}. \tag{2.226}$$

Although the system has damping, the response achieves a true steady state only when F is constant. When F is harmonic, the resulting displacement, x, also will be harmonic and reach a limit cycle as its steady state.

In all of the previous examples, the equations did not contain parameters after normalization. The solutions were therefore independent of the parameters of the problem. In this example, the damping ratio, $\xi = c/\sqrt{(mk)}$, remained. The solution is thus parameter dependent, but only on ξ. The time response for any given damping ratio is, however, independent of the parameter set.

In modeling systems where the resulting differential equation has similar terms, but with a different sign than the usual spring-mass-damper system, one can still infer something of the solution. As an example, consider the differential equation

$$\ddot{x} - \dot{x} + x = 1. \tag{2.227}$$

Because the damping term has a negative sign, one can immediately postulate (correctly) that this system has the potential to be unstable. Instead of a damped response, these systems have an unstable amplitude growth with time.

2.8.8 Other Forms of Nonlinear Equations

The differential equation for a pendulum of length, L, and mass, m, at an angle, θ, from the vertical, with damping, and an external forcing torque, T, is

$$mL^2\ddot{\theta} + c\dot{\theta} + mgl\sin(\theta) = T. \tag{2.228}$$

When normalized using the transformations

$$\begin{aligned} y &= \theta/\theta_0, \\ x &= t/\tau, \\ F &= T/T_0, \end{aligned} \tag{2.229}$$

and selecting

$$\tau = mL/c, \quad \theta_0 = m^2L^2g/c^2, \quad T_0 = mgL, \tag{2.230}$$

this system becomes

$$\ddot{y} + \dot{y} + \sin(\theta_0 y) = F. \tag{2.231}$$

The solution to this system for a given initial condition is parameter independent since all parameters have been removed from the transformed Eq. (2.231). The initial conditions are still subject to parameter dependence, and the scaling factors for both time and angle include parameters of the physical system.

The differential equation thus is seen as analogous to a spring-mass-damper system with a sinusoidal (nonlinear) elastic element and a damping ratio of unity.

2.9 COMPARISON OF MODEL AND EXPERIMENT

The application of mathematics to modeling often introduces universal numbers (e.g., universal gravitational constant, Plank's constant, speed of light in vacuo) and other variables and constants that are not universal. These system-specific constants, like the universal constants, must be determined through experiment. Often, a wealth of information and a general understanding of a phenomenon precedes the actual mathematical modeling. Galileo [18], for example, described the concept of specific gravity and buoyancy without the benefit of a symbolic algebra. Those that followed Galileo applied formalized symbols to represent in equations what Galileo had tried to express in words. The result was a mathematical model of specific gravity.

The next step in the process of formalizing a model is validating the model. It is only through continuous testing of measures of the model's performance that one can achieve a sense of assurance that the concepts and assumptions upon which the mathematical formulation is based have validity. The Bohr[14] model of the atom, although proving to be a powerful model that revolutionized our concepts of matter, eventually failed to explain all of the experimental data. Physicists are still attempting to refine their ideas and find a universal model that will define the unique construction set of subatomic particles and explain their interrelationships.

2.9.1 Concepts of a "Best" Model

The general problem of determining a mathematical model capable of representing a physical system is carried out in a repetition of two basic steps: (1) A specific structure for the model is proposed, and (2) the parameters of that structure are adjusted to maximize that model's performance. The first of these steps cannot be automated easily and depends on insight into the problem at hand and often a large measure of serendipity. However, the second step can be formulated as a strictly computational task and thereby is automated easily. Since insight and serendipity are difficult to transfer to the student, this text concentrates on the second step, the automated adjustment of model parameters to obtain the best performance possible from a specifically structured model. With automated procedures to handle the second step, the researcher's mind is free to explore new model structures without being inhibited by the tedium of step two.

---

[14] Harald Bohr (1887–1951).

2.9. Comparison of Model and Experiment

The general problem of determining the "best" parameters for models of systems often is clouded by the nonlinear relationships between the parameter values and the model's performance. Many such models are of sufficient complexity that one is unable to obtain direct expressions for the derivative of a performance function with respect to the parameter values. An additional problem associated with using derivative-based search techniques lies in the difficulty of ascertaining programming errors associated with the derivative description. Since numerical derivatives are often "noisy," one often does not even suspect that a problem exists. Because of the level of difficulty associated with model simulations, nonderivative search methods are being used more frequently to force the model's response to match the experimentally determined performance measures of the system being modeled.

The fundamental element of "the best anything," in any context, is that it is the winner, against all challengers, of a comparison test, and nothing more. If we recognize that, within the context of a certain model structure, we have a fixed, although in some instances infinite, set of challengers for the best title, we must conclude that the nature of the test, and nothing else, will determine the winner. This is an extremely important concept and cannot be overemphasized. Further, implicit in the concept of best is the idea that one of the comparands has more or less of some measurable quantity than do all of the other competing comparands. This simply means the ultimate comparison can be made by using scalar, and only scalar, representations of that exquisite quantity, regardless of the dimensionality of the parameter space that may be used to specify the model.

The choice of a measure of a model's performance is not simple and often involves an anticipation of how the nature of the measure will affect the mathematics of computing the best set of parameters. With this in mind, some frequently used measures and the mathematical mechanics they induce can be explored.

2.9.1.1 Error Measures

Generally speaking, it is often more intuitive to formulate measures of bad than of good. Within that context, the "best" has a least measure of bad. Many of our systems, and our models of them, perform over a space of independent variables such as time, flow rate, and electric potential. The system's other state variables are considered to be dependent upon (determined by) the independent variables. This immediately suggests that the performance of a model can be assessed by choosing values of the independent variables and comparing the dependent-variable values computed by the model to the values contained in the database for the system being modeled. The difference between the value of each computed dependent variable and the value for that variable in the database is an error in the model's performance. Thus we would have one error value for each dependent variable at each set of independent variables tested.

The goal is therefore to devise some scheme for converting this multiplicity of errors into a scalar representative of the total performance of the model. Ultimately, the decision must be made between one set of model parameters and another set based on which was "better." An intuitive approach would be to convert all of the errors to positive values and add them, or average them, or find the largest error, and use this result as a scalar measure of the "badness" of the model. Each of the variants of this approach satisfies our intuition. Error anywhere results in a positive measure; zero measure implies a perfect performance

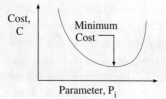

Figure 2.25. The effect of varying the model parameter on the model performance.

everywhere the model is tested. With the formulation of a method for computing a scalar measure of the performance, finding the best set of parameters for the model becomes simply the problem of finding the parameters that minimize the computed badness. If a system were to have only a single parameter, P_1, varying its value would be reflected in a change in the total error, or cost function, as depicted in Fig. 2.25.

2.9.1.2 Minimum (-Maximum)

The minimization task implied by the preceding discussion is the search over a multi-component parameter space for that vector of parameter values, P_i, that will result in the model computing a minimum scalar value for the performance measure. In the general problem, the components of the parameter space may be continuous or discrete, may be constrained to a certain small domain, or may be totally free to vary over the set of real numbers. With this in mind, the minimization problem is simply defined as follows:

> P_i^* *is the location of the minimum when* $G(P_i^*) < G(P_i^* + dP_i)$ *for all* dP_i *that produce admissible* $P_i = P_i^* + dP_i$ *and where G is the scalar objective function of the parameter vector* P_i.

In those situations in which all of the components of the parameter vector have a manageable number of admissible values (none of the parameters are continuous variables), a brute-force method is to try every possible combination. This approach becomes a usable and effective method of finding the minimum. It involves examining every idea and concept, keeping only the best, and thereby ensuring the location of the global minimum of G over the admissible space of P_i.

When some or all of the parameters have an infinite number of admissible values, the exhaustion method of finding a minimum becomes impossible to implement. For the case of all components of the parameter vector being continuous, the minimum definition remains the same but can be rephrased to reflect the continuity of the variable space:

> P_i^* *is the location of the minimum when* $G(P_i^*) < G(P_i^* + tu)$ *for all t and u that produce an admissible* $P_i = P_i^* + tu$, *where t is a scalar and u is a unit vector in* P_i.

This is still a definition of the global minimum because it implies that all admissible values are candidates for P_i^*.

If the range of t is restricted to values near zero, the above statement only tests locally and would only define a local minimum. In the context of identifying a local minimum, if continuity of the function G and its derivatives in every component of P_i is postulated, a

2.9. Comparison of Model and Experiment

local minimum of G in P_i requires that

$$G(P_i^*) < G(P_i^* + tu) \tag{2.232}$$

as $t \to 0$ for all possible directions u. This results in the necessary derivative condition

$$\left.\frac{dG}{dP_i}\right|_{P_i^*} = 0, \quad i = 1 \cdots n, \tag{2.233}$$

where n is the order of the space P_i.

This condition is symptomatic of maxima, minima, and inflection regions (saddle points). To guarantee a minima, the condition

$$\sum_{i=1}^{n}\left[\sum_{i=1}^{n}\left(u_i \frac{\partial^2 G(P_i^*)}{\partial P_i \partial P_j} u_j\right)\right] < 0 \tag{2.234}$$

is a sufficient requirement. This is equivalent to requiring that the matrix formed from the second derivatives of G with respect to the components of P_i be positive definite (all eigenvalues greater than zero).

If a point P_i^* is found that tests as a minimum using the above criteria, it can be affirmed only as a local minimum unless it is a particularly restricted class of problem. As a consequence of this intimate relation that develops between how one forms the performance measure, the nature of the model, and the methods used to find the best model parameters, some specific examples are in order.

2.9.1.3 Least Sum Square Error

Conceptually, the collection of errors, obtained by setting the independent variables of the model and system to that set of values forming the base of the physical data, is a vector of vectors. The size of this error vector can be measured in any number of computationally convenient and intuitively satisfying methods. One frequently used method introduced by Legendre[15] is the sum of the squares of the error measure. This gives a positive scalar measure that is nonzero when there is an imperfect fit between model and data anywhere over the set of independent variables. With a multiplicity of dependent variables at each point that the model is tested, the subvector of error in the dependent-variable components associated with each point in the independent variables can be sized by a sum of square errors to obtain an error vector component for that point. Normalization may be used to counter the variance in the measure of the different components of the state vector. These components, all positive, then can be summed over the independent variable set to obtain the scalar measure of the model performance.

To illustrate, consider a simple data-fitting problem where the model is a weighted linear combination of predefined functions of a single independent variable and the data are a collection of single dependent-variable values and associated independent-variable values. Thus, one may write

$$y(P_i, t) = P_1 f_1(t) + P_2 f_2(t) + \cdots + P_n f_n(t), \tag{2.235}$$

---

[15] Adrien-Marie Legendre (1752–1833).

where the f_i are predefined functions and the P_i are the weights to be determined (parameters). Note that the parameters P_i are linearly involved in the resulting model $y(P_i, t)$ as simple weights of the predefined functions f_i. In this context, the sum square error would be

$$\sum_{k=1}^{m} (y(P_i, t_k) - d_k)^2 = G(P_i), \tag{2.236}$$

where t_k, d_k are the associated independent-dependent variable pairs and m is the number of observations.

Enforcement of the zero-gradient requirement on G with respect to P_i gives a set of linear equations, the normal equations:

$$\sum_{j=1}^{n} \left[\sum_{k=1}^{m} f_i(t_k) f_j(t_k) \right] P_j = \sum_{k=1}^{m} f_i(t_k) d_k, \qquad i = 1 \cdots n. \tag{2.237}$$

If the set of functions, $f(t)$, is independent on the set of independent variable values, t_k, the system will be solvable for the vector P_i. The matrix $\sum_{k=1}^{m} f_i(t_k) f_j(t_k)$ also will be positive definite and P_i will locate the global minimum.

In the event the data, d, is an integrable function of the variable t, the sums over k become integrals over the domain of t. The problem remains the same otherwise.

Our problem of finding the minimum has been reduced to the solving of a set of linear equations, a straightforward computational task. It must be pointed out that the choice of which functions should be included in the set f is up to the analyst.

Weighted Square Error
It is common to have reason to weigh the error at certain points in the independent domain more heavily than the error elsewhere. To accommodate this, a positive weighting function is introduced, either $\omega(t)$ for the integrable case or w_k for the summed error measure. The only modification to the equations is the introduction of $\omega(t)$ into the integrals on t or w_k into the sums on k. The sum version becomes

$$\sum_{k=1}^{m} \omega_k (y(P_i, t_k) - d_k)^2 = G(P_i), \tag{2.238}$$

$$\sum_{j=1}^{n} \left[\sum_{k=1}^{m} \omega_k f_i(t_k) f_j(t_k) \right] P_j = \sum_{k=1}^{m} f_i(t_k) \omega_k d_k, \qquad i = 1 \cdots n. \tag{2.239}$$

Orthogonal Functions
So far, the discussion has considered the number of functions in the model to be finite. The best fit requires the solving of a set of n linear equations, where n is the number of functions used in the linear mix. Certain sets of functions have an infinite number of members; for example, $\sin(iqt)$, where i is any positive integer. The use of such a function set would result in an infinite number of simultaneous equations to solve; specifically, for the integrable case,

$$\sum_{j=1}^{\infty} \left[\int \omega(t) f_i(t) f_j(t) dt \right] P_j = \int f_i(t) \omega(t) d(t) dt, \qquad i = 1 \cdots \infty. \tag{2.240}$$

This infinite set is not solvable unless it has a special structure. One workable structure occurs when the functions are orthogonal to each other over the domain of t with the weighting function $\omega(t)$. In this case the matrix of the equations is diagonal and each equation is solvable independent from all of the others. For example, the functions $\sin(iqt)$, $i = 1 \cdots \infty$, are orthogonal, with weighting function $\omega(t) = 1.0$, over any interval $a < t < a + \pi/q$. This means the terms $\int \omega(t) f_i(t) f_j(t) dt = 0$ for all $i \neq j$. Use of the harmonic functions, $\sin(iqt)$ and $\cos(iqt)$, in this manner results in the P_i becoming the coefficients of a Fourier series. A Fourier series is thus a least sum square error fitting of the harmonic functions to the data regardless of how many terms are used and each coefficient is independent from all others. One simply adds terms until a sufficient accuracy is achieved.

When the specificity of a problem dictates the use of functions that are not inherently orthogonal, the selection of a proper error weighting function, $\omega(t)$, may allow one to reduce the system of equations to a diagonal form and solve the set for any order. Note that the error weighting function is chosen solely to allow an easy solution to the normal equations and may not enhance the fit of the model functions to the data when n is small. Therefore, when the model has a small function base, error weight functions based on facilitating the computation of the function multipliers may not give the "best" results.

2.9.2 Other Natural Error Measures

If one considers the same dependent-independent state variable description of our physical system and multicomponent error described previously, two other measures with high intuitive content are often used to convert the error to a scalar. One of these is the average weighted absolute error

$$G(P_i) = \left(\frac{1}{m}\right) \sum_{k=1}^{m} \omega_k |y(P_i, t_k) - d_k|. \tag{2.241}$$

The other is the maximum weighted absolute error

$$G(x) = \text{MAX}[\omega_k |y(P_i, t_k) - d_k|] \cdots, \quad k = 1 \cdots m. \tag{2.242}$$

Both of these are simple to compute and easy to interpret but their derivatives are not practical to compute, regardless of the form of the variation of y with respect to P_i. As a consequence, these measures are seldom used to develop analytical solutions analogous to the linear least sum square error problem solutions.

2.9.3 Nonlinear Problems

Models with linearly involved parameters are important, but the behavior that such models can describe is restricted in scope and form. As soon as models of most real problems are expanded to include real effects, they become nonlinear. For example, the model $y(P_i, t) = P_i \sin(Qt)$ is linear in P_i, but as soon as one considers Q to be one of the design parameters, $y(P, t) = P_1 \sin(P_2 t)$, is nonlinear in the parameter space, P_i. As the physical behavior of the system increases in complexity, it becomes more likely that a nonlinear model will be required to capture the essence of the behavior. With a nonlinearity in the parameter-model relation, the normal equations for the least sum square error problem are nonlinear. Depending upon the level of complexity, the difficulty in computing the normal equation is often more trouble than it is worth. With the introduction of models that may not have defined derivatives, the normal equations are not even defined.

Generally, the nonlinear problems are approached by using computer methods that traverse the space of parameter values, P_i, and only sample the function $G(P_i)$ occasionally. Press et al. [36], discuss this class of methods in *Numerical Recipes*.

Postulates are made regarding the continuity of $G(P_i)$ to justify extrapolating and interpolating the behavior of G about the sampled points. A typical assumption is that, locally, G expands as a Taylor[16] series in all directions:

$$G(t) = G(P_i^*) + \left(\sum_{i=1}^n G_i^* u_i\right) t + \left(\sum_{j=1}^n \sum_{i=1}^n u_j G_{ji}^* u_i\right) \frac{t^2}{2} + O(t^3), \qquad (2.243)$$

where G_i^* is the ith component of the gradient of G with respect to P_i and evaluated at P_i^*; G_{ji}^* is the ijth component of the Jacobian of the gradient evaluated at P_i^*, called the Hessian of G; and u_i is the ith component of a unit direction vector. The parameter t is a dummy parameter. This assumes that the function G is at least quadratic in P_i locally.

Most search techniques evaluate the function, and perhaps derivatives as well, at a point, determine a direction, and then proceed in that direction to a new point at which G is less than at the starting point. The requirement that the value of G at the new point be less than at the old point forces stability on the algorithm. Ultimately, a point P_i^* is found where no direction can be found to walk to decrease G. This point is taken to be a local minimum.

2.9.3.1 Gradient Method

If the gradient of the function G is available, the choice of direction exactly counter to the direction of the gradient produces a maximum rate of decrease in G for a unit step. This is the embodiment of the method known as *the method of steepest descent*. The direction so computed will decrease G so that there will exist some step size in that direction that will result in a new P_i that produces G less than before. Since the method finds a local minimum where the values of the components of the gradient vanish, most implementations of the method suffer in the end game where determination of the walking direction and the step size become uncertain. Further, for nonquadratic functions, this method tends to be inefficient in finding the general location of a relative minimum.

If the second derivatives of G are also available, the end game can be greatly improved by the Gauss–Newton method; the application of the Newton–Raphson method to finding the P_i that forces the gradient vector to vanish.

2.9.3.2 Nongradient Method

The computation of a gradient is often an onerous task even when the gradient is defined. For a broad class of problem, the gradient is not defined or is difficult to compute.

---

[16] Brook Taylor (1685–1731).

2.9. Comparison of Model and Experiment

The calculation of a difference approximation to the gradient is generally expensive in a computational sense and subject to considerable error. Computational experimentation with properly formulated nongradient search methods shows them to converge better and more efficiently than the best gradient methods using differenced gradient approximations. There is a twofold motivation for using nongradient search methods: They work better when gradients are not analytical and they require less programming in all cases.

As a class, these multidimensional search methods are called *direction set methods*. They all involve use of a complete set of search directions in the space P_i, the members of the set having special relations to one another. The naive set would be the coordinate directions of the space P_i. They also use a cycle of successive searching along each direction of the set as a fundamental unit of the method. The method presented here in some detail is a modification of the basic quadratically convergent method of Powell [35][17].

We postulate a continuous space P_i and a continuous objective function $G(P_i)$, perhaps defined as above by some computed measure of performance of our model. We further assume the availability of an algorithm capable of searching along a direction vector in P_i until a relative minimum of G is found along that line.

The algorithm is initialized by choosing the coordinate directions as the original direction set,

$$S_i = a_i P_i, \quad i = 1 \cdots n, \tag{2.244}$$

where a_i is a heuristically determined scaling factor, and P_i is the ith coordinate vector.

The second step of the initialization is a search from a given starting point, P_0, along S_n until a relative minimum is found:

$$P_0 \leftarrow P_0 + t S_n, \tag{2.245}$$

with t such that $dG(P_0 + tS_n)/dt = 0$. This point, P_0, is taken as the origin of the first search cycle. Note that S_n is tangent to an isovalue contour of G in P.

A search cycle then consists of three separate phases:

Phase 1. Successive searches are made along each direction in the direction set

$$P_i \leftarrow P_{i-1} + t S_i \quad \text{for } i = 1 \cdots n, \tag{2.246}$$

with t such that $dG(P_{i-1} + tS_i)/dt = 0$. This locates the point P_n.

Phase 2. A new search vector, $S_{n+1} = P_n - P_0$, is formed and a search is made to a new starting point for the next cycle,

$$P_0 \leftarrow P_n + t S_{n+1}, \tag{2.247}$$

with t such that $dG(P_n + tS_{n+1})/dt = 0$. This is often referred to as an acceleration search since phase 1 frequently detects a valley in the G versus P surface. Note that S_n was tangent to an isovalue contour at both P_0 and P_n and this fact is essential to S_{n+1} being a good direction for changing G.

---

[17] M.J.D. Powell (1936–).

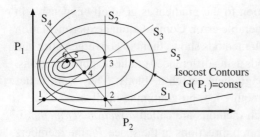

Figure 2.26. Several search cycles for Powell's method.

Phase 3. Before commencing the next search cycle, S_{n+1} is compared to each of the other vectors in the direction set. The vector most aligned with S_{n+1} is cast from the set and those with higher index are rippled down:

$$S_i \leftarrow S_{i+1} \quad \text{for} \quad i = k \cdots n, \tag{2.248}$$

where k is the index of the cast-out vector. Note also that S_n is again tangent to an isovalue contour at P_0. The search algorithm then goes back to Phase 1 for the next cycle.

Convergence is indicated when the decrease of the function G from the beginning of a cycle to the end of that cycle is less than a set limit, or when the change in P_i^* is less than a set limit. A generally pessimistic view of convergence is adopted and the algorithm is restarted at least twice before convergence is declared.

Figure 2.26 depicts several cycles of the algorithm in a two-dimensional problem. Although *Numerical Recipes* [36] somewhat diminishes the role of the routine that determines the location of the minimum in a search along a direction vector, all of the calls for function evaluation come from that routine and any inefficiencies there can negate good performance by the executive routine. The FORTRAN optimization subroutine MNWD4B is available through File Transfer Protocol (FTP) at *ftp://ftp.me.unm.edu/pub/mnwd4b.for* and is an implementation of the algorithm described here. To ensure maximum performance, it has the relative minimum routine integrated into the direction-set search routine. Further, the direction-set routine does not use unit vectors but scales the direction vectors up and down so that the relative minimum routine expects the minimum to occur near the dummy parameter value of 1.0. The relative minimum routine begins each search with a quadratic interpolator but switches to a cubic as soon as a minimum has been indicated by positive curvature of the quadratic. In this fashion this relative minimum routine has an average of six to eight function evaluations for every relative minimum detected. This statistic was established over a test period spanning several years and hundreds of different problems.

2.9.4 Constrained Problems

In the previous discussions, it has been implicit that the vector space P was continuous and open in all of the components, P_i. This generally is not true for real problems. Almost all real problems have some type of interval constraint on the parameters. There will be limits beyond which solutions just do not make sense regardless of the computed value of a performance measure. Further, often there will be some intermediate computed quantity that must satisfy some constraint in order that the rest of the model performance be computable. For example, should the inverse cosine of a quantity $q(P)$ be required for the model to simulate the physical system, the quantity q must satisfy $-1 \leq q \leq +1$. If q does not satisfy this constraint, the mathematics break down and P, the trial value of the parameter vector, must be considered infeasible. This means there is a region in P that must be avoided by the search routine. To further complicate matters, the boundary between the feasible and infeasible may not be simply defined and in fact may require much more analysis to define than the solution to the original problem. For this reason, and other economies (laziness), a more simpleminded approach to constraint enforcement is indicated so that an unconstrained search algorithm can be used to effect the search for the minimum.

2.9.4.1 Variable Mapping

Occasionally problems occur in which a finite region can be defined as the feasible region of a variable. In such cases an unconstrained variable can be mapped from an infinite region into the finite region. For example, $a = \sin(x)$ maps $-\infty < x < +\infty$ onto $-1 \leq a \leq +1$. With appropriate mapping functions, most a priori defined domain constraints can be seen as mappings from unconstrained domains. In this fashion it becomes impossible to get outside the feasible domain. All search walks will remain interior to the feasible region. With the use of nonderivative search methods, these mappings do not impose a heavy programming burden.

2.9.4.2 Penalty Functions

In complex problems one does not usually have nicely defined feasible-infeasible boundaries. The usual case is that, halfway through a complex simulation computation, an intermediate variable takes on a mathematically forbidden value and further computation is impossible. If this happens at only one point of the independent variable set, the model parameters must be considered infeasible. Other types of infeasibility may not be as dramatic, a stress may exceed a limit, and computation of the model's performance can proceed to completion in spite of the fact that the limit has been violated. Although it may seem a contradiction, in the following discussion some sets of infeasible model parameters will be considered more infeasible than other sets of infeasible parameters. This will be necessary in order to use an exuberant search method that would otherwise have the tendency to tramp across a feasibility boundary in the heat of the chase.

The search program MNWD4B uses a relative minimum search that assumes a quadratic or cubic form for its interpolation model. This suggests that the computed error, objective function, or model performance indicator should have at least continuity of function and derivative if any performance efficiency is expected during the search process. Thus, if an infeasible model parameter set is presented to be evaluated by computing the model performance, the performance computed must not result in a discontinuity in the performance function. This will require some special considerations since an infeasible set has no meaningful performance.

The fundamental concept is simply that every boundary, when approached from the feasible side, will be numerically recognizable before the boundary is crossed. For example, if the square root of a number must be taken to evaluate a formula, one simply examines that number and determines the proximity from above to 0.0. When the computed value of the number that will violate the boundary is closer than a preset distance from the boundary, a penalty function is computed and added to the model performance. This penalty

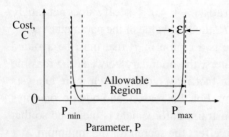

Figure 2.27. Cost penalty function.

function is formed with zero value and derivative at the precursor to the boundary and with a boundary value greater than or equal to the greatest feasible model performance. Should the the number be beyond the boundary, the number is clipped to the boundary to allow completion of the performance calculation without a discontinuity.

The functional form of the penalty function is up to the analyst, but considerable success has been had using a quadratic form. This functional behavior is shown in Fig. 2.27. The imposition of the quadratic form has been mechanized in the form of the FORTRAN subroutine TPT (Test, Penalize, Truncate), available through FTP at *ftp://ftp.me.unm.edu/pub/tpt.for.*

The logic of this quadratic penalty program is as follows:

P is computed and is to be constrained;

Lower boundary P_{min};

if $(P_{min} - P > \epsilon)$, then $C \leftarrow C + (MFV/\epsilon^2)(P_{min} - P)^2$;

if $(P < P_{min})$, then $P \leftarrow P_{min}$;

Upper boundary P_{max};

if $(P_{min} - P < \epsilon)$, then $C \leftarrow C + (MFV/\epsilon^2)(P_{min} - P)^2$;

if $(P > P_{min})$, then $P \leftarrow P_{min}$;

P is then used in subsequent calculations.

Note when P is beyond the boundary, the penalty function, C, is computed before the truncation of P to the boundary. The accumulation implies that the penalty is zeroed before each performance evaluation.

The imposition of the maximum feasible value, MFV, on the penalty function at the boundary ensures that the search algorithm will converge inside the feasible region. The adding of the penalty for each boundary violation has the effect of making some parameter sets more infeasible than others; specifically, when there are several points in the independent-variable set, the model parameter sets that indicate infeasibility at a large number of points will have greater indicated performance error than model parameter sets that indicate infeasibility at just a few points.

The determination of the maximum feasible value of the error or other bad performance indicator is problem specific and will require some analysis but often will be obtained quite simply. The value of the parameter ϵ will be constraint specific and can be adjusted to get closer to a constraint boundary with successive searches with the same search program. In many real problems, the minimum parameter set will lie at or on a boundary. See Fiacco and McCormick [16] for additional details.

2.10 SUMMARY

In this chapter, many tools have been presented to assist in model formulation, including

1. methods of model formulation;
2. vector and tensor notation for symbolic mathematical representation of variables in models;
3. a formalized normalization technique for testing the importance of various terms in the governing equations of the model;
4. a method for optimization of model variables to experimental data; if this match proves successful, the resultant model can be said to be validated within the assumptions of the model.

Perhaps the most important material presented in this chapter is the normalization method. This technique allows the modeler to accomplish three basic goals:

1. Establishing a unit-domain for all problem variables.
2. Determining the relative importance of the terms in the differential equation of the model. This is a natural consequence of the normalization process, though the selection of the scaling functions may have to be altered until the proper characterizations are found.
3. Gaining new physical insights into the system under study. This insight is derived from the scaling variables, which provide characteristic lengths for the domains of the problem. Frequently, in design problems, the exact solution is not important, but the relationships that characterize the domains yield the design constraints for the system under study.

In linear systems and, more rarely, in nonlinear systems, it is common for the normalization process to produce solutions that appear to be parameter independent. Thus, the infinite family of solutions one normally encounters, with various parameters defining where in this family the solution appears, falls away to a single solution. This apparent discrepancy is explained when one realizes that the functions we use for the transformation of coordinates (which we have termed scaling here) are dependent on the parameters of the system model. This dependency is usually in terms of groups of the important parameters; the same groups one would normally get from a dimensional analysis and the Buckingham pi theorem. Thus, as the parameters change, the physical domain is stretched or compressed, thereby yielding solutions that appear identical in the scaled domain.

Though we have used linear transformations almost exclusively, the whole idea of normalization of the domain sometimes encourages the use of nonlinear transformations. This technique was presented in Example 2.9, where a problem with an infinite domain was reduced to one with finite domain. The insight gleaned from this transformation was important, with the added bonus that the problem became analytically tenable in the transformed domain.

The next four chapters seek to present methods for the solution of the models that are derived using these tools. Chapter 3 introduces the first methods one might employ; Chapter 4 introduces numerical techniques; Chapters 5 and 6 introduce approximation solution methods.

2.11 PROBLEMS

2.1 An equation giving the velocity of an airplane on a runway after landing with a velocity V_0 is

$$\frac{W}{g}\frac{dv}{dt} = -F - kV^2 \qquad v(0) = V_0,$$

where: W = weight of airplane,
g = gravitational constant,
F = frictional force from brakes,
k = air resistance coefficient.

The frictional force model for the brakes is

$$F = C_b W,$$

where C_b is the frictional coefficient between the tires and the ground.

(1) Determine the normalization variables and derive relations for the constants A, B, Y_0 in the normalized form of the equation,

$$\frac{dy}{dx} + Ay^2 = B, \qquad y(x=0) = Y_0.$$

(2) Sketch what you believe the solution should look like. Be certain to annotate (scale) your axes and indicate if the airplane will come to a complete stop.

(3) From your analysis, find the numeric values of A, B, Y_0 using the following numerical data:

$g = 32.2$ ft-lbm/lbf-s^2,
$k = 0.30$ lbf-sec^2/ft^2,
$W = 185{,}000$ lbm,
$C_b = 0.095$ lbf/lbm,
$V_{0\min} = 250$ mph,
$V_{0\max} = 350$ mph.

2.2 Modify the normalization of the temperature-dependent resistance problem presented in Example 2.7 by assuming that

$$\theta = (T - T_0)/(T_m - T_0), \qquad (2.249)$$
$$I = I/I_0, \qquad (2.250)$$
$$y = t/\tau, \qquad (2.251)$$

where T_0, T_m are the initial and maximum temperatures, respectively, and τ is the characteristic timescale of the problem.

2.3 Extend the analysis of the temperature-dependent resistance given in Example 2.7 by assuming that $\theta = 1$. Obtain an analytical expression relating the normalized variables of current and time. Then, if $R_p \ll 1$, show that $I = e^{-y}$ and obtain an analytic solution for $\theta(y)$.

2.11. Problems

Figure 2.28. Circular arc of angle θ.

2.4 In railroad surveying work, it is common practice to replace the length of a circular arc (S) by its chord (C) as shown in Fig. 2.28. If we define the measurement error as

$$\epsilon = \frac{S - C}{S},$$

how big may θ be if the error is to be 0.01 or less?

2.5 Consider the problem of cooling of the Earth as described in Problem 8.2. After dropping negligible terms, the governing differential equation for the surface temperature of the Earth, T_{es}, can be shown to be

$$\left[\frac{4mC_p k}{\sigma h}\right] \frac{dT_{es}}{dt} - T_{es} + \mathcal{F}\frac{T_s^4}{T_{es}^3} = 0.$$

Normalize the above equation and comment on the time and temperature scales that arise in the normalization process. Attempt to sketch the solution.

2.6 Consider the following simultaneous linearized differential equations for the variation of normalized temperature (θ) and current (I) with respect to normalized time (y):

$$\frac{d\theta}{dy} = 2kI + (2k-1)\theta + (1-3k), \qquad \theta(0) = 1,$$

$$\frac{dI}{dy} = -2(k-2)\theta + (1-6k)I + 3(2k+1), \qquad I(0) = 1.$$

You may recognize this as the temperature-dependent resistance problem of Example 2.7.

(1) Using a symbolic manipulation language, find the exact solution for $I(y)$ and $\theta(y)$.

(2) For time steps of 0.1, 0.5, and 1.0 numerically solve for θ and I over a time range of $0 < y < 5$. For this analysis, the product of the electrical and thermal time constants and power factors is $k = R_{\text{pwr}}, R_\tau = 2.0$. You may use any numerical solution technique discussed in class.

(3) Using the computer, plot your results as θ and I vs y from both the numerical solution and the exact solution.

(4) Discuss your results, explaining any differences in results between step sizes and comment on their comparison to the exact solution.

2.7 Consider the following differential equation describing the change in position, $r(t)$, of a small body of mass m in the gravitational field of a larger body of mass M and radius R:

$$m\frac{d^2r}{dt^2} = -\frac{mM\gamma}{r^2} = -\frac{mk}{r^2}, \qquad (2.252)$$

subject to the initial conditions $r(0) = r_0$, $\dot{r}(0) = 0$.

(1) First, linearize and then normalize this differential equation, and solve it for the dimensionless position, y, as a function of dimensionless time, x. Use the scaling

TABLE 2.2. Physical Symbols and Data for the Source and Body

Parameter	Value, units
Heat transfer flux rate, $\dot{q}$	Btu/hr-ft²
Nominal source temperature, T_s	10,400°R
Nominal body temperature, T_b	520°R

transformations $y = (r - r_0)/(R - r_0)$ and $x = t/\tau$. The resulting differential equation should be of the form

$$\ddot{y} = y - A, \qquad y(0) = \dot{y}(0) = 0. \qquad (2.253)$$

Show your work and specify the constants A and τ to achieve the proper normalization of the system.

(2) Using a symbolic algebra program, find the exact solution for $y(x)$.

(3) Using the computer, plot your results as y versus x for both the linearized solution and the exact solution (see text). Use the following data for your solutions:

Initial position, $r_0 = 2.00002 \times 10^7$ ft,
Radius of Earth, $R = 2.0 \times 10^7$ ft,
Acceleration of gravity, $g = 32.2$ ft/s².

(4) Discuss your results.

2.8 The radiation heat transfer flux per unit time and area from a source to a body, $\dot{q}(T_s, T_b)$, is given by Eq. (2.254). You are to linearize this mathematical function with respect to both the source temperature, T_s, and the body temperature, T_b:

$$\dot{q} = \sigma \left(T_s^4 - T_b^4 \right). \qquad (2.254)$$

Use the data shown in Table 2.2 and plot the ratio of the linearized to actual heat flux for the cases

$$T_s = 10,400°R : T_b = [400..1000°R], \qquad (2.255)$$
$$T_b = 520°R : T_s = [10,000..11,000°R]. \qquad (2.256)$$

2.9 A growing phenomenon with finite resources to grow upon will ultimately reach a value beyond which it cannot go. A differential equation to express this is

$$\frac{dx}{dt} = k(A - x)x, \qquad x(0) = X_0,$$

where the magnitude x of the growing phenomenon at time t has a limiting value of A, an initial value of X_0, and k is a physical constant. You are to normalize the given equation, and discuss the meaning of the characteristic scales you use for the normalization of the x and t domains.

CHAPTER 3

Exact Solution Methods

The examples of complex real-world nonlinear problems having exact solutions are very few in number. It is, of course, always advantageous to obtain an analytic solution, if at all possible, because of the wealth of information that is contained in an analytic formula. All of the trends and the nuances of the solution are contained in the analytical result. It is appropriate, then, to review some of the classical exact solution techniques that produce analytical solutions. This effort also serves the dual purposes of giving instruction in exact solution methods and providing experience in the behavior of nonlinear systems. Many nonlinear problems that are not readily solvable in an exact manner may be simplified to a form for which a solution might be found. Examples of some of these also are included in various sections within this chapter.

As one progresses through this chapter, it is important to note that the techniques and methods that are presented can be converted readily to algorithms on computers, using ordinary programming languages, simulation languages, or symbolic algebra systems. Because this translation is so machine- and language-specific and the field of computing is evolving so rapidly, no attempt is made here to present code or algorithms. The tools provided on modern computing platforms frequently include symbolic algebra systems, and these are of extreme value in assisting the reader through the material in this chapter.

3.1 TECHNIQUE OF LINEARIZATION

Mathematical models of real engineering systems typically are very involved and have many nonlinear terms. The nonlinearities often make the resulting system of governing equations difficult or impossible to solve exactly. In this section, the approach taken is to assume that, as a first approximation, the nonlinear terms within the differential equation may be simplified and approximated by a simple linear relationship. From a mathematical viewpoint, we must restrict this simplification to a specific domain within which the nonlinearity is approximated with a linear relationship. This concept is depicted in Fig. 3.1, where the nonlinear function is represented as a straight line over the domain $x[A, B]$. As noted in Fig. 3.1, the function is only closely matched over the desired domain. If the function is derived from experimental data, it is recommended that the original data be approximated using the method of least squares to fit the data in the region of interest with a linear equation: $f(x) = a + bx$. In other cases, the function to be linearized is a mathematical representation of the behavior of a nonlinear function.

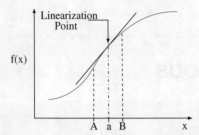

Figure 3.1. Function linearization over the domain $x[A, B]$.

3.1.1 Method of Finding Linear Coefficients

Mathematicians have been representing arbitrary functions using a broad range of analytic relationships for many years. One of the more common representations is the use of a polynomial series. This was originally published in 1715 by Brook Taylor and is described in his work *Methodus Incrementorum* [49]. In this representation, one begins with some known nonlinear function, $f(x)$. If this function is analytic over the domain, that is, if both the function and its derivatives exist and are continuous over the domain, then a Taylor expansion of the function about the midpoint of the domain, a, may be written

$$f(x) = f_a + \left.\frac{df}{dx}\right|_a (x-a) + \left.\frac{d^2 f}{dx^2}\right|_a \frac{(x-a)^2}{2!} + \left.\frac{d^3 f}{dx^3}\right|_a \frac{(x-a)^3}{3!}$$

$$+ \cdots + \left.\frac{d^n f}{dx^n}\right|_a \frac{(x-a)^n}{n!} + \cdots + R_n. \tag{3.1}$$

Since all of the terms beyond the first-order term represent nonlinear terms in x, they are dropped to obtain the linearized approximation

$$f(x) \approx f(a) + \frac{df}{dx}(a)(x-a). \tag{3.2}$$

Using this method, several examples of nonlinear expressions have been linearized and the results are given in Table 3.1. It is left as an exercise to the reader to verify these approximations. In the following examples, the method just described is used to linearize some simple functions. The first example is a linear representation of the sine function.

TABLE 3.1. Linear Approximations

Item	Expression	Linear approximation	Next term in series
1	$1/(1 \pm m)$	$1 \mp m$	$0.5m^2$
2	$(1 \pm m)^n$	$1 \pm nm$	$n(n-1)m^2/2$
3	e^m	$1 + m$	$0.5m^2$
4	$\ln(1 + m)$	m	$-m/2$
5	$\sin(m)$	m	$-(m^3)/6$
6	$\cos(m)$	1	$-0.5m^2$
7	$(1 + m_1)(1 + m_2)$	$1 + m_1 + m_2$	$m_1 m_2$
8	$\sin^{-1}(m)$	m	$(m^3)/6$
9	$\sinh(m)$	m	$(m^3)/6$
10	$\cosh(m)$	1	$0.5m^2$
11	$\tanh(m)$	m	$-(m^3)/3$

3.1. Technique of Linearization

EXAMPLE 3.1: LINEARIZED TRIGONOMETRIC FUNCTIONS 1

Trigonometric substitutions have been used previously to linearize equations. Using Taylor expansions for sine and cosine yields

$$\sin(x_0 + x) = \sin x_0 + \left.\frac{d(\sin x)}{dx}\right|_{x_0} x + \left.\frac{d^2(\sin x)}{dx^2}\right|_{x_0} \frac{x^2}{2!} \cdots$$

$$= \sin x_0 + x \cos x_0 - \frac{x^2}{2} \sin x_0 \cdots.$$

Similarly, applying Taylor series to the function $\cos(x_0 + x)$ yields

$$\cos(x_0 + x) = \cos x_0 - x \sin x_0 - \frac{x^2}{2} \cos x_0. \tag{3.3}$$

If one drops terms of order higher than one, the linear approximations for these functions are obtained:

$$\sin(x_0 + x) = \sin x_0 + x \cos x_0, \tag{3.4}$$

$$\cos(x_0 + x) = \cos x_0 - x \sin x_0. \tag{3.5}$$

The following example is a linearization of the arctan( ) trigonometric function.

EXAMPLE 3.2: LINEARIZED TRIGONOMETRIC FUNCTIONS 2

Can $\arctan(x)$ be expanded about $x = 0$? The use of a Taylor-series expansion yields the proper result:

$$\tan^{-1}(x) = \tan^{-1} x_0 + \frac{d(\tan^{-1} x)}{dx} x + \cdots$$

$$= 0 + x + 0 - \frac{x^3}{3} + \cdots.$$

As a final example, the nonlinear effects of gravity are linearized over a specific domain.

EXAMPLE 3.3: GRAVITY EFFECTS

The nonlinear effects of gravity are not apparent over small variations in radius of separation between gravitational bodies. In this example, the general expression for gravity,

$$F(R) = \frac{\gamma m M}{R^2}, \tag{3.6}$$

is linearized around the point $R = R_0$. This is done using a Taylor expansion of the above equation:

$$F(R) = F(R_0) + \left.\frac{dF}{dR}\right|_{R_0} (R - R_0) + \left.\frac{d^2 F}{dR^2}\right|_{R_0} \frac{(R - R_0)^2}{2}$$

$$+ \cdots + \left.\frac{d^n F}{dR^n}\right|_{R_0} \frac{(R - R_0)^n}{n!} + \cdots. \tag{3.7}$$

Thus, evaluating the derivatives of Eq. (3.6) and inserting them into Eq. (3.7) yields

$$F(R) = \frac{\gamma m M}{R_0^2} - 2\frac{\gamma m M}{R_0^3}(R - R_0) + 3\frac{\gamma m M}{R_0^4}(R - R_0)^2 + \text{H.O.T.} \tag{3.8}$$

Neglecting all of the $(R - R_0)^n$ terms with $n > 1$ results in the linearized equation

$$F(R) \approx \frac{\gamma m M}{R_0^2}\left[3 - 2\frac{R}{R_0}\right]. \tag{3.9}$$

3.1.2 Linearization of a Multivariable Function

The following procedure depicts the representation of a function with a large number of independent variables. This method is valid only if the solution function is analytic. As before, the procedure is to expand the function about some operating point in a Taylor series and then discard all but the linear and constant terms.

If the function is multivariate in x_i, that is, $f(x_1, x_2, \ldots, x_n) = f(x_i)$, then the Taylor series expanded about the point $x_i = a_i$ is

$$f(x_i) = f(a_i) + \sum_{i=1}^{n} \left.\frac{\partial f}{\partial x_i}\right|_{x=a}(x_i - a_i) + \frac{1}{2}\sum_{i=1}^{n}\left.\frac{\partial^2 f}{\partial x_i^2}\right|_{x=a}(x_i - a_i)^2$$
$$+ \frac{1}{2}\sum_{i=1}^{n}\sum_{j=1}^{n}\left.\frac{\partial^2 f}{\partial x_i \partial x_j}\right|_{x=a}(x_i - a_i)(x_j - a_j)(1 - \delta_{ij}) + \cdots + \mathcal{R}. \tag{3.10}$$

The function δ_{ij} is the Kronecker delta function and has values of one for all $i = j$ and zero for all $i \neq j$. In this series, only the constant and linear terms are retained, resulting in

$$f(x_i) = f(a_i) + \sum_{i=1}^{n}\left.\frac{\partial f}{\partial x_i}\right|_{x_i=a_i}(x_i - a_i). \tag{3.11}$$

To demonstrate the application of this method, the behavior of a nonlinear flow regulator follows.

EXAMPLE 3.4: FLOW REGULATOR

Consider a flow regulator as shown in Fig. 3.2. For such flows, it is possible to write an equation for the flow, q, through the regulator having a variable area, A, and pressure drop, P, in the following manner:

$$q = C_D A\sqrt{P} = q(A, P), \tag{3.12}$$

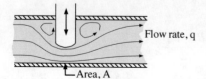

Figure 3.2. Simple flow regulator system.

3.1. Technique of Linearization

where C_D is the discharge coefficient for the valve. Applying the technique of linearization yields the expression [42]

$$q(A, P) = q(A_0, P_0) + \left.\frac{\partial q}{\partial A}\right|_{A_0, P_0} (A - A_0) + \left.\frac{\partial q}{\partial P}\right|_{A_0, P_0} (P - P_0) + R. \tag{3.13}$$

The two derivatives are the influence, or sensitivity coefficients, for the area and pressure effects, respectively. The terms A_0, P_0 represent the area and pressure at some known operating value. This equation reduces to its linear form,

$$q(A, P) = C_D A_0 \sqrt{P_0} \left[1 + \left(\frac{A}{A_0} - 1\right) + \frac{1}{2}\left(\frac{P}{P_0} - 1\right)\right]. \tag{3.14}$$

This equation is seen to be linear with respect to both A and P. The limitation on this approximation, based on the next nonzero-term-neglected method, can be shown to be

$$\left|\frac{P - P_0}{4P_0}\right| \ll 1. \tag{3.15}$$

Interestingly, this limitation does not involve the area, A, because the flow rate was already linear with respect to the area. Equation (3.15) can be reduced further to

$$|(P/P_0) - 1| \ll 4, \tag{3.16}$$

which states that the variation from the reference pressure, P_0, must be kept well below four times the reference pressure.

In the next example, the use of linearization is applied in the development of a mathematical model of a flyball governor.

EXAMPLE 3.5: MODEL OF A FLYBALL GOVERNOR

Consider the use of a flyball governor to regulate the speed, ω, of a steam engine. The flyball system is shown in Fig. 3.3. The net rotational force (NRF) in a direction perpendicular to the swing arm is thus

$$\text{NRF} = m\omega^2 l \sin\phi \cos\phi - mg \sin\phi, \tag{3.17}$$

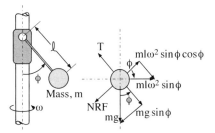

Figure 3.3. Flyball governor system.

and the system is in equilibrium when NRF $= 0$. Thus, at some equilibrium state where $\omega = \omega_0$ and $\phi = \phi_0$, the following relationship exists:

$$\cos \phi_0 = \frac{g}{l\omega_0^2}. \tag{3.18}$$

Consider the motion of the system when the dynamic terms become important. It is of value to define

gravity torque $= -mgl \sin \phi$,

centrifugal torque $= m\omega^2 l^2 \sin \phi \cos \phi$,

damping torque $= -bl(d\phi/dt)$.

The net torque about the swing arm is therefore available for rotational acceleration, or

$$ml^2 \frac{d^2\phi}{dt^2} = m\omega^2 l^2 \sin \phi \cos \phi - mgl \sin \phi - bl\frac{d\phi}{dt}. \tag{3.19}$$

Let us simplify this equation and set $l = 1$ for convenience. This yields

$$m\frac{d^2\phi}{dt^2} = m\omega^2 \sin \phi \cos \phi - mg \sin \phi - b\frac{d\phi}{dt}. \tag{3.20}$$

Letting the entire rotational inertia be represented by I, there are two principal torques in effect:

$T_s =$ torque due to steam engine (tends to increase ω),

$T_d =$ constant torque required to drive the load (tends to decrease ω).

This may be written as

$$I\frac{d\omega}{dt} = T_s - T_d. \tag{3.21}$$

The next equation is a simplified model in its own right; it attempts to relate the angle of the governor to the torque developed by the steam engine. The engine torque, T_s, is assumed to be linearly related to the steam flow rate, which in turn is assumed to be harmonically related to the valve angle, ϕ. These assumptions are included in Eq. (3.22):

$$T_s = T_l + k [\cos \phi - \cos \phi_0], \tag{3.22}$$

where $T_l =$ steam torque at set point, $k =$ gain of governor, $T_s =$ steam torque, and $\phi_0 =$ governor angle at equilibrium set point (ω_0). Thus,

$$m\frac{d^2\phi}{dt^2} = m\omega^2 \sin \phi \cos \phi - mg \sin \phi - b\frac{d\phi}{dt}, \tag{3.23}$$

$$I\frac{d\omega}{dt} = k \cos \phi - T, \tag{3.24}$$

where $T = T_d - T_l + k \cos \phi$. At equilibrium, $\omega = \omega_0$ and $\phi = \phi_0$, which consequently

3.1. Technique of Linearization

results in the statement that all changes in ϕ or ω vanish along with their second-order derivatives:

$$\frac{d\phi}{dt} = \frac{d^2\phi}{dt^2} = \frac{d\omega}{dt} = 0. \tag{3.25}$$

Therefore, Eq. (3.21) may be solved for the equilibrium relationship $\cos\phi_0 = T/k$, or

$$\omega_0^2 = \frac{g}{\cos\phi_0} = \frac{gk}{T}. \tag{3.26}$$

The procedure is thus to linearize around the equilibrium values of the governor angle, ϕ_0, and the engine speed, ω_0, by choosing $\phi = \phi_0 + y$ and $\omega = \omega_0 + x$. The formal derivatives of the above are thus

$$\frac{d\phi}{dt} = \frac{d(\phi_0 + y)}{dt} = \frac{dy}{dt}, \tag{3.27}$$

$$\frac{d^2\phi}{dt^2} = \frac{d^2 y}{dt^2}, \tag{3.28}$$

$$\frac{d\omega}{dt} = \frac{d(\omega_0 + x)}{dt} = \frac{dx}{dt}. \tag{3.29}$$

It is also assumed that the variations (x, y) are small and apply the following rules:

- Since x and y are small quantities, only those terms that contain x or y will be retained; no higher powers or cross products will be kept:

$$(\omega_0 + x)^2 = \omega_0^2 + 2\omega_0 x + x^2 \approx \omega_0^2 + 2\omega_0 x. \tag{3.30}$$

- Since y is very small, it is further assumed that

$$\sin y = y,$$
$$\cos y = 1,$$
$$\sin\phi_0 + y = \sin\phi_0 \cos y + \cos\phi_0 \sin y$$
$$= \sin\phi_0 + y\cos\phi_0,$$
$$\cos\phi_0 + y = \cos\phi_0 \cos y - \sin\phi_0 \sin y$$
$$= \cos\phi_0 - y\sin\phi_0.$$

The application of these rules results in

$$m\frac{d^2 y}{dt^2} = m\left(\omega_0^2 + 2\omega_0 x\right)(\sin\phi_0 + y\cos\phi_0)(\cos\phi_0 - y\sin\phi_0)$$
$$- mg(\sin\phi_0 + y\cos\phi_0) - b\frac{dy}{dt}, \tag{3.31}$$

$$I\frac{dx}{dt} = k(\cos\phi_0 - y\sin\phi_0) - T. \tag{3.32}$$

By neglecting the higher-order terms, the following equations are obtained:

$$m\frac{d^2 y}{dt^2} = m\omega_0^2(\cos^2\phi_0 - \sin^2\phi_0)y + 2mx\omega_0 \sin\phi_0 \cos\phi_0$$
$$- mgy\cos\phi_0 - b\frac{dy}{dt}, \tag{3.33}$$

$$I\frac{dx}{dt} = -ky\sin\phi_0. \tag{3.34}$$

A single third-order equation may be obtained by combining these two equations:

$$m\frac{d^3y}{dt^3} + b\frac{d^2y}{dt^2} + \left(\frac{mg\sin^2\phi_0}{\cos\phi_0}\right)\frac{dy}{dt} + \left(\frac{2mgk\sin^2\phi_0}{I\omega_0}\right)y = 0. \tag{3.35}$$

The characteristic equation is

$$mL^3 + bL^2 + \frac{mg\sin^2\phi_0}{\cos\phi_0}L + \frac{2mgk\sin^2\phi_0}{I\omega_0} = 0. \tag{3.36}$$

3.1.3 Taylor-Series Expansion Errors

The application of a Taylor series to linearize a nonlinear function, of necessity, must introduce errors because of the omitted terms. The magnitudes of the omitted terms can be thought of as a measure of the size of the error of linearization. There are two methods generally employed to estimate this error, the *classical* and the *first-term-neglected* (FTN) methods.

3.1.3.1 Classical

Details on this method can be found in any book on advanced calculus. It is the most common method used to determine the error in truncating a Taylor series. If one simplifies the Taylor series using the substitution

$$f(x) = f(a) + \left.\frac{df}{dx}\right|_a (x-a) + R, \tag{3.37}$$

then the remainder may be estimated by the upper bound of R,

$$R = \frac{(x-a)^2}{2!}\left.\frac{d^2f}{dx^2}\right|_{a+c(x-a)}, \tag{3.38}$$

where one selects the value of c within the range $0 \le c \le 1$ to obtain the maximum value for R. As an example of this procedure, consider the function $\sin x$. This function is to be expanded in a Taylor series about the point $x = x_0$ and linearized. This results in the linear function (with respect to δ)

$$\sin(x_0 + \delta) = \sin x_0 + \delta \cos x_0 + R, \tag{3.39}$$

with the residual error

$$R = -\left[\frac{\delta^2}{2!}\right]\sin(x_0 + c\delta), \tag{3.40}$$

or, by properly choosing the value of c, this reduces to

$$R \le -\left[\frac{\delta^2}{2!}\right]. \tag{3.41}$$

3.1.3.2 First-Term-Neglected Method

If the Taylor series is written as an expansion about the point $x = a$, the result is

$$f(x) = f(a) + \left.\frac{df}{dx}\right|_{x=a} (x - a) + \eta, \tag{3.42}$$

where η is the next nonzero term in the Taylor expansion. The series is a reasonably good approximation if the first truncated term is much less than the second term:

$$\left.\frac{df}{dx}\right|_{x=a} (x - a) \ll \eta. \tag{3.43}$$

This inequality relationship for the independent variable x defines a radius of convergence of the Taylor series representation of $f(x)$. The reader might wish to attempt both previous methods on Taylor series for $\tan^{-1} x$.

To demonstrate the use of this technique, an example is presented. As in the preceding example, this is combined with the development of a model. The model presented next is a representation of the handling of an automobile. As you read through the development of this model, look for its intended purpose, expert knowledge, and constraints.

EXAMPLE 3.6: LINEARIZATION OF AUTOMOBILE HANDLING PROBLEM

In Section 2.5.8, the following differential equation describing the response of an automotive vehicle was obtained:

$$I\frac{d\omega}{dt} = aF_F - bF_R, \tag{3.44}$$

where F_F and F_R are the cornering forces on the front and rear tires, respectively. For small slip angles, it is assumed that

$$F_F = C_F \alpha_F, \qquad F_R = C_R \alpha_R, \tag{3.45}$$

where C_F and C_R are termed the front and rear cornering power. The slip angle also may be represented in terms of the yaw angle, ω:

$$\alpha_R = \tan^{-1}\left(\frac{-b\omega}{V}\right) \tag{3.46}$$

and

$$\alpha_F = \tan^{-1}\left(\frac{a\omega}{V}\right) - \delta. \tag{3.47}$$

The equation of motion becomes

$$I\frac{d\omega}{dt} = aC_F[\tan^{-1}(a\omega/V) - \delta] - bC_R[\tan^{-1}(-b\omega/V)]. \tag{3.48}$$

This equation is still nonlinear and is approximated by expanding $\tan^{-1}$ about the point $x_0 = 0$, or

$$\tan^{-1} x = x - x^3/3 + x^5/5. \tag{3.49}$$

Based on Table 3.1 and the results of preceding sections, the first term, x, may be used to approximate $\tan^{-1}(x)$ if $x \gg x^3/3$ or $|x| \ll \sqrt{3}$. Therefore $\tan^{-1}(a\omega/V)$ is replaced by $(a\omega/V)$, $\tan^{-1}(-b\omega/V)$ by $(-b\omega/V)$ as long as $|(a\omega/V)| \ll \sqrt{3}$ or $|(a\omega/V)| \ll \sqrt{3}$. As a numerical note, if one uses practical values such as $a = 5$ ft, $v = 30$ mph, linearization is reasonable provided $|\omega| \ll 10.7$ rad/s. This is a reasonable criterion readily accommodated by everyday drivers, and a linear equation is finally obtained:

$$I\frac{d\omega}{dt} = \frac{(a^2 C_F + b^2 C_R)\omega}{V} - aC_F\delta. \tag{3.50}$$

EXAMPLE 3.7: GRAVITY EFFECTS

The linearization of gravity effects described earlier in Example 3.1 carries with it an associated error. By using the FTN method, it is possible to place a bound on the range over which Eq. (3.9) may be applied. The first term neglected must be much smaller than the linear term that was retained, or

$$\left|3\frac{\gamma mM}{R_0^4}(R-R_0)^2\right| \ll \left|2\frac{\gamma mM}{R_0^3}(R-R_0)\right|. \tag{3.51}$$

This can be reduced to the inequality constraint

$$\left|\frac{R}{R_0} - 1\right| \ll \frac{2}{3}. \tag{3.52}$$

3.2 DIRECT INTEGRATION

For any given linear or nonlinear differential equation or system of equations, the ideal solution would be an analytic expression or expressions that exactly satisfy the governing equation(s) and all of the boundary or constraint conditions. However, all problems do not submit to exact solution. A very small class of nonlinear problems may be directly integrated, and the general techniques one may use to accomplish this are studied here.

If the independent and dependent variables can be manipulated so that they are in separate terms, the resultant equations are assured of being integrable. This is the classical technique of separation of variables. In the following sections, examples of this technique and other forms of direct integration are demonstrated.

3.2.1 Radiatively Heated Thermal Capacitance

Let us consider the problem of a radiatively heated thermal capacitance. The amount of radiant energy emitted by a surface is modeled by Stefan's law as proportional to the fourth power of the absolute temperature of a surface. The differential relationship describing

3.2. Direct Integration

the rate of change of stored thermal energy as a balance of incoming and outgoing thermal radiant energy is

$$C\frac{dT}{dt} + kT^4 = kT_i^4, \tag{3.53}$$

where k is termed Stefan's constant. The independent and dependent variables in this equation may be separated into two terms,

$$\frac{dT}{T_i^4 - T^4} = \frac{k}{C}dt, \tag{3.54}$$

or, using $\xi = T/T_i$, the differential equation becomes

$$\frac{d\xi}{1 - \xi^4} = \frac{k}{C}T_i^3\, dt. \tag{3.55}$$

This equation may be directly integrated to obtain

$$\frac{1}{4}\ln\frac{T_i + T}{T_i - T} + \frac{1}{2}\tan^{-1}(T/T_i) = \frac{k}{C}T_i^3 t + D. \tag{3.56}$$

This is an exact solution to the problem as formulated. It satisfies the linearized governing differential equation but no initial condition has been imposed. The constant D is determined when the initial condition is specified.

EXAMPLE 3.8: NUMERICAL EXAMPLE

This exact solution affords us an excellent opportunity to evaluate the effects of our technique of linearization. To simplify the mathematics, let us define a new dimensionless time variable, τ, as

$$\tau = (k/C)T_i^3 t. \tag{3.57}$$

To solve for D, the initial condition of $T(0) = T_i/2$ is imposed on the exact solution of Eq. (3.56):

$$D = 0.25\ln(3) + 0.5\tan^{-1}(0.5) = 0.50647. \tag{3.58}$$

The exact solution for the linearized case can be obtained readily:

$$0.25\ln\frac{T_i + T}{T_i - T} + 0.5\tan^{-1}(T/T_i) = \tau + 0.50647. \tag{3.59}$$

A graphical comparison[1] of the solutions of the linearized and the original nonlinear problems is depicted in Fig. 3.4.

The question of the range of applicability of the solution in this instance is complicated by the convergence of the linearized solution to the exact solution. The linearization, however, was accomplished at a specific value of temperature, $T/T_i = 1.0$. Thus, for small values of τ (a region near the initial condition), the linearized function for T^4 will be acceptable and the solution will be close to the exact value.

---

[1] The student might attempt to perform the linearization and solution of this problem and comparison as an exercise.

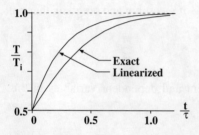

Figure 3.4. Exact, linearized solutions for a radiatively heated thermal capacitance.

3.2.2 Spring-Dashpot System

Consider the following equations representing a massless element suspended on a spring and a dashpot that act in parallel. The total force acting on the system, F, is

$$F = F_D + F_S, \tag{3.60}$$

and the differential equation describing the response, x, with respect to time, t, and the initial condition is given by

$$a\frac{dx}{dt} + bx^2 = F, \qquad x(0) = 0. \tag{3.61}$$

If we let $A = b/a$ and $B = F/a$, the equation is simplified to

$$\frac{dx}{dt} + Ax^2 = B. \tag{3.62}$$

Separating variables and integrating yields

$$\frac{1}{\sqrt{AB}} \tanh^{-1}\left(x\sqrt{A/B}\right) = t + C. \tag{3.63}$$

Requiring that the solution satisfy the initial condition yields the result[2]

$$x = \sqrt{B/A}\, \tanh\left(\sqrt{AB}\,t\right). \tag{3.64}$$

This result may be plotted (see Fig. 3.5) to show the response, x, over time, t. Let us consider the following linear differential equation which is similar to Eq. (3.62):

$$\frac{dx}{dt} + Ax = B, \qquad x(0) = 0. \tag{3.65}$$

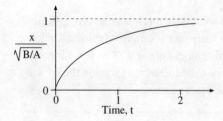

Figure 3.5. Unforced nonlinear system with square-law spring, linear dashpot.

---

[2] It would be a valuable exercise for the student to compare the exact and linearized solutions for this problem and describe the limitations for his linearized solution.

3.2. Direct Integration

This equation may be readily integrated to obtain

$$x = (B/A)[1 - e^{-At}]. \tag{3.66}$$

Note that the dynamic response of a linear system depends only upon the differential equation and its parameters, not on the forcing function. The dynamics of nonlinear systems, however, depend upon the system's differential equation, its parameters, and the forcing function.

3.2.3 Automobile Handling Revisited

The automobile handling problem was defined previously and its governing differential equation was developed in Section 2.5.8:

$$I\frac{d\omega}{dt} = [aC_F]\alpha_F - [bC_R]\alpha_R = aF_F + bF_R. \tag{3.67}$$

Based on an estimate of the errors from the linearization of the tire forces, this equation is only valid for small angles, $\alpha < 2°$. To obtain a more realistic model requires a closer evaluation of the relationship between the forces exerted by the tires and their slip angles. A simple approach to this problem is to fit a polynomial to the actual experimentally measured force–slip-angle data. This produces a more accurate model, even though it is not phenomenologically based. The polynomial takes on the form

$$F_F = C_F[\alpha_F + a_2\alpha_F^2 + a_3\alpha_F^3 + \cdots a_n\alpha_F^n + \cdots], \tag{3.68}$$

$$F_R = C_R[\alpha_R + b_2\alpha_R^2 + b_3\alpha_R^3 + \cdots b_n\alpha_R^n + \cdots]. \tag{3.69}$$

This results in a more realistic differential equation:

$$I\frac{dr}{dt} = aC_F\left(\alpha_F + a_2\alpha_F^2 + a_3\alpha_F^3 + \cdots + a_n\alpha_F^n + \cdots\right)$$
$$- bC_R\left(\alpha_R + b_2\alpha_R^2 + b_3\alpha_R^3 + \cdots + b_n\alpha_R^n + \cdots\right). \tag{3.70}$$

Our previous work related the slip angles to the slew velocity linearly:

$$\alpha_F = a\omega/V - \delta \quad \text{and} \quad \alpha_R = -b\omega/V. \tag{3.71}$$

Changing from ω as a dependent variable to α_F is done by using the substitution $\omega = (\alpha_F + \delta)V/a$, yielding

$$\frac{d\omega}{dt} = \frac{V}{a}\left[\frac{d\alpha_F}{dt} + \frac{d\delta}{dt}\right], \tag{3.72}$$

$$\alpha_R = -(\alpha_F + \delta)b/a. \tag{3.73}$$

After substitution into the differential equations and truncating terms beyond the squared terms, we obtain

$$\frac{IV}{a}\left[\frac{d\alpha_F}{dt} + \frac{d\delta}{dt}\right] = aC_F(\alpha_F + a_2\alpha_F^2) + bC_R\left[\left(\frac{b}{a}\right)(\alpha_F + \delta) - \left(\frac{b}{a}\right)^2 b_2(\alpha_F + \delta)^2\right]. \tag{3.74}$$

This is a very difficult nonlinear differential equation to solve. As a simplification, let us consider a maneuver in which the car is sliding but the steering angle is steady and held at neutral, or

$$\delta = \frac{d\delta}{dt} = 0. \tag{3.75}$$

Although this is not a realistic situation, it may give us insight into the solution of the general problem. This simplification results in the modified differential equation

$$\frac{d\alpha_F}{dt} + A_1\alpha_F + A_2\alpha_F^2 = 0, \tag{3.76}$$

where the coefficients A_1 and A_2 are

$$A_1 = -\frac{(a^2 C_F + b^2 C_R)}{IV}, \tag{3.77}$$

$$A_2 = -\frac{\left(a^2 a_2 C_F - \frac{b^3 b_2 C_R}{a}\right)}{IV}. \tag{3.78}$$

If we solve Eq. (3.76) subject to the initial condition $\alpha_F(0) = \alpha_{F_0}$, we obtain

$$\alpha_F = \frac{A_1 \alpha_{F_0}}{A_1 \exp(A_1 t) + A_2 \alpha_{F_0}[\exp(A_1 t) - 1]}. \tag{3.79}$$

From the linear solution, we previously obtained the result

$$\alpha_F = \alpha_{F_0} \exp(-A_1 t). \tag{3.80}$$

By comparing these two solutions, it is obvious that the nonlinear solution contains information not available in the linear solution. This new information provides added insight into how the parameters of the problem influence the system's behavior. This enables the modeler to adjust parameter values, thereby enhancing the performance of the suspension. Thus, the exercise of assuming the unrealistic constant slip angle did provide useful insight into the problem by bringing to light information not previously known.

3.2.4 Gravitational Attraction

Consider the physical system in Fig. 3.6, representing the gravitational attraction of two planetary bodies. This is a simple problem governed by the inverse square law for the forces on interplanetary bodies. If the universal gravitational constant is Υ and the local acceleration of gravity is g, then we may model the mutual attractive force as

$$F = -mM\Upsilon/r^2. \tag{3.81}$$

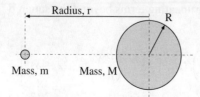

Figure 3.6. Simple gravitational attraction problem.

At the surface of the larger mass, the force is

$$F = -mg = -mM\Upsilon/R^2 \tag{3.82}$$

or

$$M\Upsilon = gR^2 = k, \tag{3.83}$$

and we obtain

$$F = -mk/r^2. \tag{3.84}$$

The governing equation for the motion of the smaller body is therefore

$$m\frac{d^2r}{dt^2} = -\frac{mk}{r^2}. \tag{3.85}$$

If we define $v = dr/dt$, then the governing equation becomes

$$v\frac{dv}{dr} = -k/r^2. \tag{3.86}$$

This may be integrated, and the initial conditions applied to obtain

$$v^2 = 2k\left[\frac{1}{r} - \frac{1}{r_0}\right], \tag{3.87}$$

where $v(0) = 0$, $r(0) = r_0$. This may be solved algebraically for r:

$$\frac{dr}{dt} = v = \pm\sqrt{2gR^2\left[\frac{1}{r} - \frac{1}{r_0}\right]}. \tag{3.88}$$

This may be integrated to obtain

$$t = \frac{\left[\sqrt{rr_0^2 - r^2 r_0} + \frac{\pi}{4}r_0^2 + \frac{1}{2}r_0^{3/2}\arcsin\left(1 - 2\frac{r}{r_0}\right)\right]}{\sqrt{2gR^2}}. \tag{3.89}$$

EXAMPLE 3.9: GRAVITY ACTING ON A BALL

A numerical example is in order. If $g = 32.2$ ft/s^2, $R = 20,000,000$ ft, $r_0 = 20,002,000$ ft (representing a fall from 2000 ft), then the time for r to diminish from r_0 to R may be computed as 11.15 s. It is left as an exercise for the student to compare this result to the classical case where the acceleration $d^2y/dt^2 = -g =$ constant.

3.3 VARIATION OF PARAMETERS

Another exact-solution technique involves finding a solution through the complementary solution to a problem. To better understand this, we first look at linear systems and then extend the method to nonlinear problems.

If the complementary solution function for a linear differential equation is known, the method of variation of parameters can be used to find the particular integral. The method is based on the observation that the particular integral of a linear equation may be written as the product of one or more parts of the complementary function and one or more functions of the independent variable. An example in which this technique is used will help in understanding the basic concept. Consider a simple linear, first-order, ordinary differential equation,

$$\frac{dx}{dt} + x = 1, \qquad x(0) = 0. \tag{3.90}$$

The form of the solution is assumed to be a product of two functions,

$$x(t) = u(t)v(t), \tag{3.91}$$

where $u(t)$ is the solution to the homogeneous form of the differential equation

$$du/dt + u = 0, \tag{3.92}$$

which has the solution $u = e^{-t}$. At this point the constant of integration has been suppressed. The solution function and the differential equation are thus

$$x(t) = e^{-t}v(t), \tag{3.93}$$

$$e^{-t}\left[\left(\frac{dv}{dt} - v\right) + v\right] = 1. \tag{3.94}$$

Upon integration of the differential equation, the unknown function, $v(t)$, is found to be $v = A + \exp(t)$, and the full solution, $x = uv$ is

$$x(t) = e^{-t}[A + e^t] = 1 + Ae^{-t}. \tag{3.95}$$

Applying the initial conditions results in the final solution,

$$x(t) = 1 - e^{-t}. \tag{3.96}$$

For nonlinear problems, the same procedure is utilized, but the complementary function is found from the linear equation obtained by suppressing all of the nonlinear terms. To demonstrate this approach, the automobile handling problem is considered in the next example.

EXAMPLE 3.10: AUTOMOBILE HANDLING PROBLEM

The governing differential and initial condition for the simplified automobile handling problem is

$$\frac{d\alpha_F}{dt} + A_1\alpha_F + A_2\alpha_F^2 = 0, \qquad \alpha_F(0) = \alpha_{F_0}. \tag{3.97}$$

If the α_F^2 term is suppressed and α_F is defined as $\alpha_F(t) = u(t)v(t)$, where $u(t)$ is the solution of the homogeneous differential equation, then

$$\frac{du}{dt} + A_1 u = 0 \tag{3.98}$$

and the solution function $u(t) = e^{-A_1 t}$ is obtained. This is reintroduced into the differential equation to obtain

$$e^{-A_1 t}\left[\frac{dv}{dt} - A_1 v + A_1 v + A_2 v^2 e^{-A_1 t}\right] = 0 \tag{3.99}$$

or

$$\frac{dv}{dt} + A_2 v^2 e^{A_1 t} = 0. \tag{3.100}$$

This may be separated and integrated to obtain

$$v(t) = \frac{1}{\left[C - \frac{A_2}{A_1} e^{-A_1 t}\right]}, \tag{3.101}$$

or, remembering that $\alpha_F = uv$, one may obtain

$$\alpha_F(t) = \frac{e^{-A_1 t}}{\left[C - \frac{A_2}{A_1} e^{-A_1 t}\right]}. \tag{3.102}$$

The constant C may be evaluated by applying the initial condition, $\alpha_F(0) = \alpha_{F_0}$, which yields the solution

$$\alpha_F(t) = A_1 \alpha_{F_0} / \left[A_1 e^{A_1 t} + A_2 \alpha_{F_0}(e^{-A_1 t} - 1)\right]. \tag{3.103}$$

It is left to the student to try $d\alpha_F/dt + A_1 \alpha_F + A_2 \alpha_F^2 = 1$, where $\alpha_F(0) = 0$.

This technique does not always proceed to a solution. In some cases, the second step of finding a solution for the unknown function, $v(t)$, does not lend itself to exact solution.

3.4 EQUATIONS LEADING TO ELLIPTIC INTEGRALS

The question of finding general solutions to differential equation models of nonlinear systems is simple: there are no general methods. Rather, one has to seek solutions by whatever methods are available for various classes of systems. One such classification includes equations whose solutions contain elliptic integrals. The general form of the governing differential equation is

$$\left(\frac{dx}{dt}\right)^2 = \sum_{i=0}^{n} a_i x^i, \tag{3.104}$$

where n is an integer. These types of equations have three groups of solutions according to the value of the exponent, n, as

Exponential for $n = 0, 1, 2$,
Elliptic functions if $n = 3, 4$,
Hyperelliptic if $n > 5$.

In this section, only the elliptic forms of Eq. (3.104) are studied and solutions are derived. The final form of the solution is not a closed analytical form, but instead is a functional,

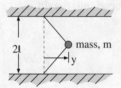

Figure 3.7. Suspended-mass problem.

integral representation of the solution. A numerical computation of the functional then can be computed to attain the solution. As in previous sections, the technique is developed through the use of example problems.

3.4.1 Suspended Mass

The motion of a suspended mass, depicted in Fig. 3.7, is described mathematically by a nonlinear equation. In this system, the mass, m, oscillates in a horizontal plane, a motion only possible if the string restraining the mass changes in length. The length at any time may be shown to be

$$L(y) = \sqrt{l^2 + y^2}. \tag{3.105}$$

In this example, we assume that there is no applied force on this system, rendering the equations homogeneous. This implies that the initial conditions must be non-homogeneous if the solution is to be nontrivial:

$$m\ddot{y} + 2EA\frac{y}{l} + 2(T - AE)\frac{y/l}{\sqrt{1 + (y/l)^2}} = 0. \tag{3.106}$$

Through the following substitution, the governing differential equation can be linearized:

$$\frac{y}{\sqrt{l^2 + y^2}} = \frac{y/l}{\sqrt{1 + (y/l)^2}} = \frac{y}{l}\left[1 - \frac{(y/l)^2}{2} + \frac{3(y/l)^4}{8} + \cdots\right]. \tag{3.107}$$

By substitutions of $a = 2T/ml$, $b = (AE - T)/ml^3$, and $v = dy/dt$, the following expanded differential equation is obtained:

$$v\frac{dv}{dy} + ay + by^3 = 0. \tag{3.108}$$

The variables in this equation may be separated and the equation integrated to obtain

$$v^2 = a(Y_0^2 - y^2) + \frac{b}{2}(Y_0^4 - y^4). \tag{3.109}$$

This solution is of the form

$$\left(\frac{dy}{dt}\right)^2 = \sum_{i=0}^{4} a_i y^i \tag{3.110}$$

or

$$\frac{dy}{dt} = \pm\sqrt{a(Y_0^2 - y^2) + \frac{b}{2}(Y_0^4 - y^4)}. \tag{3.111}$$

3.4. Equations Leading to Elliptic Integrals

The positive root in this equation is not physically realistic over the first half-cycle, and thus it is possible to again separate the variables and obtain

$$dt = \frac{-dy}{\sqrt{a(Y_0^2 - y^2) + \frac{b}{2}(Y_0^4 - y^4)}}, \tag{3.112}$$

which may be integrated to give

$$t = -\frac{1}{\sqrt{a}} \int_{Y_0}^{y} \frac{dy}{\sqrt{(Y_0^2 - y^2)(1 + \frac{b}{2a}(Y_0^2 + y^2))}}. \tag{3.113}$$

This integral may be simplified by the nonlinear transformation

$$y = Y_0 \cos \phi, \tag{3.114}$$

$$dy = -Y_0 \sin \phi \, d\phi \tag{3.115}$$

$$= -Y_0 \sqrt{1 - (y/Y_0)^2} \, d\phi \tag{3.116}$$

$$= -\sqrt{Y_0^2 - y^2} \, d\phi. \tag{3.117}$$

This may be inserted into the integral and, using $k^2 = bY_0^2/2(a + bY_0^2)$, the following is obtained:

$$t = \frac{1}{\sqrt{a + bY_0^2}} \int_0^{\phi} \frac{d\phi}{\sqrt{1 - k^2 \sin^2 \phi}}. \tag{3.118}$$

This is called an elliptic integral, and it has been evaluated numerically and *tabulated* with independent parameters k and ϕ. Symbolically, the solution function is represented as

$$t = \frac{1}{\sqrt{a + bY_0^2}} \mathcal{F}(k, \phi), \tag{3.119}$$

where $\mathcal{F}(k, \phi)$ is an elliptic integral. Note that $\mathcal{F}(k, \pi/2)$ is termed a complete elliptic integral [19].

EXAMPLE 3.11: NUMERICAL EVALUATION

Consider the special case given by the selection of constants $a = 10$, $b = 100$, $Y_0 = 1$. Also, remembering that $y = Y_0 \cos(\phi)$, this results in the solution

$y =$	Y_0	0	$-Y_0$	0	Y_0	... etc.
$\phi =$	0	$\pi/2$	π	$3\pi/2$	2π	

We evaluate the time for a one-quarter period and then multiply by four to obtain the full period:

$$t_{\text{cycle}} = \frac{4}{\sqrt{a + bY_0^2}} \int_0^{\pi/2} \frac{d\phi}{\sqrt{1 - k^2 \sin^2 \phi}}$$

$$= \frac{4}{\sqrt{a + bY_0^2}} \mathcal{F}\left(k, \frac{\pi}{2}\right),$$

but for this example, we may numerically compute this using

$$k = \sqrt{\frac{bY_0^2}{2(a+bY_0^2)}} = \sqrt{\frac{100}{2(10+100)}} = 0.674. \tag{3.120}$$

Therefore,

$$t_{\text{cycle}} = \frac{4}{\sqrt{110}} \mathcal{F}(0.674, \pi/2)$$

$$= \frac{4(1.82)}{10.488}$$

$$= 0.6941 \text{ s}.$$

The number 1.82, above, was taken from a table of complete elliptic integrals [2].

As an interesting exercise, we consider a linear, homogeneous simplification of Example 3.11. Our goal is to determine if the nonlinearity changes the length of the period of the problem. The governing equation without the nonlinearity is then

$$\ddot{y} + 10y = 0. \tag{3.121}$$

This problem has the simple solution

$$t_{\text{cycle}} = \frac{2\pi}{\sqrt{10}} = 1.987 \text{ s}. \tag{3.122}$$

Comparing these two results, we see that as the nonlinearity of a problems increases, this can have an important effect upon the system response. The period of the nonlinear solution also varies with the magnitude of the initial condition, whereas the period of the linear problem does not. This is a common behavior of nonlinear systems.

If one studies the time history of this problem using the given parameters, it is found that $t_{\text{cycle}} = \mathcal{F}(0.674, \phi)/10.49$. Mathematicians have tabulated this function, $\mathcal{F}(\alpha, \phi)$ where, for this example, $\alpha = \sin^{-1}(k) = 42.5°$ (see Table 3.2). This permits the computation of the period for this system as $t_{\text{cycle}} = \mathcal{F}(42.5, \phi)/10.49$. Plotting these results as y versus time yields Fig. 3.8.

3.4.2 Simple Pendulum

The equation of motion of a simple pendulum is given as

$$\ddot{y} + (g/l)\sin y = 0, \tag{3.123}$$

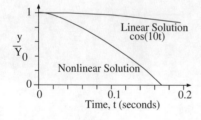

Figure 3.8. Linearized vs exact solution obtained using elliptic integrals.

3.4. Equations Leading to Elliptic Integrals

TABLE 3.2. (α, ϕ) vs ϕ, Elapsed Time

ϕ	$y = \cos \phi$	$\mathcal{F}(42.5, \phi)$	Time, t
0	1.000	0.000	0.000
10	0.985	0.175	0.017
20	0.940	0.352	0.034
30	0.866	0.534	0.051
40	0.766	0.724	0.069
50	0.643	0.923	0.088
60	0.500	1.133	0.108
70	0.342	1.353	0.129
80	0.174	1.584	0.151
90	0.000	1.820	0.174

where $v = \dot{y}$ and $y(0) = Y_0$, $v(0) = 0$. This reduces to

$$v = -\sqrt{\frac{2g}{l}[\cos y - \cos Y_0]},$$

$$= -\sqrt{\frac{4g}{l}[\sin^2(Y_0/2) - \sin^2(y/2)]}.$$

Letting $k = \sin(Y_0/2)$ and $\sin(y/2) = k \sin(\phi)$, then the period of the pendulum becomes

$$t_{\text{cycle}} = 4\sqrt{\frac{l}{g}} \int_0^{\pi/2} \frac{d\phi}{\sqrt{1 - k^2 \sin^2 \phi}} \qquad (3.124)$$

$$= 4\sqrt{\frac{l}{g}} \mathcal{F}\left(k, \frac{\pi}{2}\right). \qquad (3.125)$$

The obvious question raised here is whether one can generalize this approach or not. To evaluate this question, consider the differential equation

$$\ddot{y} + ay + by^3 + cy^5 = 0. \qquad (3.126)$$

Through simple extrapolation of previous work, the following result was obtained:

$$t_{\text{cycle}} = \text{const } \mathcal{G}(a, b, c). \qquad (3.127)$$

The function $\mathcal{G}$ cannot be readily tabulated since a three-dimensional table would be required.

Tabular values for simple elliptic integrals can be found in Abramowitz and Stegun [2]. Sometimes these functions are referred to as Legendre's canonical incomplete elliptic integrals of the first, second, and third kind. The original tabular values were compiled by Legendre and reprinted in 1934 [26]. The introduction to the reprint includes equations for "computing" the elliptic integrals. At the time of this work, this was necessarily done manually by human "computers." Most current symbolic algebra systems contain intrinsic or library functions for the computation of elliptic integrals.

3.5 POWER-SERIES METHOD

The most general approach to obtaining a solution to a model of a nonlinear system is to assume that the solution takes on the form of a power series in the independent variable. There is no formal proof that the resulting solution will be unique, that it is a stable solution, or that the finite representation has a useful region of convergence to the actual solution. As in other sections, this concept is developed through the use of examples.

3.5.1 Spring-Damper Problem

Consider a mathematical model of the classical nonlinear spring-damper problem with a nonlinear second-order spring. The governing equation for this system and the initial condition are

$$\frac{dx}{dt} + Ax^2 = B, \qquad x(0) = 0. \tag{3.128}$$

The second-order term, Ax^2, is nothing more than a simplistic model of the nonlinear stiffness of the elastic element in the system. The solution is assumed to be of the form $x(t) = \sum_{i=0}^{N}(C_i t^i)$. This solution also must satisfy the initial conditions; thus $C_0 = 0$ and, taking only fifth-order approximations, the solution is

$$x = \sum_{i=1}^{5}(C_i t^i), \tag{3.129}$$

which is inserted into the differential equation to obtain

$$\frac{dx}{dt} = C_1 + 2C_2 t + 3C_3 t^2 + 4C_4 t^3 + 5C_5 t^4$$
$$= B - Ax^2$$
$$= B - A(C_1 t + C_2 t^2 + C_3 t^3 + C_4 t^4 + C_5 t^5)^2.$$

Collecting terms with like powers yields

$$F(t) = (C_1 - B) + (2C_2)t + (3C_3 + AC_1^2)t^2 + (4C_4 + 2C_1 C_2 A)t^3$$
$$+ (5C_5 + 2C_1 C_3 A + C_2 A)t^4 + \cdots = 0.$$

It is readily seen that this equation must be zero at $t = 0$. In fact, it is zero for all other values and has zero derivatives. Thus, it follows that

$$\begin{aligned}
C_1 - B &= 0 \Longrightarrow C_1 = B \\
2C_2 &= 0 \Longrightarrow C_2 = 0 \\
3C_3 + AC_1^2 &= 0 \Longrightarrow C_3 = -AB^2/3 \\
4C_4 + 2C_1 C_2 A &= 0 \Longrightarrow C_4 = 0 \\
5C_5 + 2C_1 C_3 A + C_2^2 A &= 0 \Longrightarrow C_5 = 2A^2 B^3/15 \\
&= 0 \Longrightarrow C_6 = 0 \\
\vdots \quad \vdots \quad &\Longrightarrow C_n.
\end{aligned} \tag{3.130}$$

Therefore, we obtain

$$x = Bt - AB^2 t^3/3 + 2A^2 B^3 t^5/15 - \cdots \tag{3.131}$$

or

$$x = \sqrt{B/A}\left[(AB)^{1/2}t - \tfrac{1}{3}(AB)^{3/2}t^3 + \tfrac{2}{15}(AB)^{5/2}t^5 - \cdots\right]. \tag{3.132}$$

The exact solution for this problem may be shown to be

$$x = \sqrt{B/A}\,\tanh\left[\sqrt{AB}\,t\right]. \tag{3.133}$$

This corresponds exactly to the first three terms in our assumed solution. However, the power series converges only for

$$t < \sqrt{\frac{5}{2AB}}. \tag{3.134}$$

As an exercise, you might try to solve the following equation:

$$\ddot{y} + ay + by^3 = 0, \qquad Y(0) = 1, \qquad \dot{Y}(0) = 0, \tag{3.135}$$

using the data $a = 10$ and $b = 100$, and compare it to the exact solution obtained using elliptic integrals. Use only a two- and three-term power-series expansion.

3.6 PICARD'S METHOD

In Section 3.5, it was mentioned that the power-series method did not provide assurances that the solution obtained was unique or that it converged to the exact solution, provided a sufficiently large number of terms was included. The method presented here is of enormous value in that such a uniqueness and convergence proof does exist. These proofs are presented by Lefschetz[3] [25, 26] but only the method is presented here.

This method, attributed to Picard,[4] uses an iterative solution analytical procedure. The basic technique uses a fixed mathematical procedure in which each successive application of the procedure improves on the solution obtained. Consider the following differential equation:

$$\frac{dx}{dt} = f(x, t), \qquad x(t_0) = X_0. \tag{3.136}$$

Integration of this equation between the initial condition, $x(t_0) = X_0$, and any subsequent value of x at an arbitrary time, t, yields

$$x = X_0 + \int_{t_0}^{t} f(x, z)\, dz. \tag{3.137}$$

We cannot evaluate the integral since it requires full knowledge of the solution, x, which is the value of that integral. We can, as a first approximation, evaluate the integral with x set at the initial condition, or $f(x(0), t)$ is $f(X_0, t)$. Thus, to a first approximation,

$$x_1 = X_0 + \int_{t_0}^{t} f(X_0, z)\, dz. \tag{3.138}$$

---

[3] Solomon Lefschetz (1884–1972).
[4] Charles Emile Picard (1856–1941).

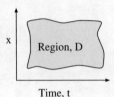

Figure 3.9. Problem domain for Picard's method.

Here the subscripts on x represent the number of steps taken to improve the approximation. Similarly, we can evaluate the integral repetitively as

$$x_n = X_0 + \int_{t_0}^{t} f(x_{n-1}, z) \, dz. \tag{3.139}$$

Consider the domain shown in Fig. 3.9. It can be guaranteed [7] that if $f(x, t)$ and df/dt are continuous and finite in the region D, then f is an analytic function, and there is a unique solution to the integral of Eq. (3.139) within D, and Picard's method will find it.

The use of the digital computer provides tools for applying this method to a level not previously attainable. Symbolic algebraic manipulation programs generally provide automatic integration features which may be iteratively applied and terms grouped. It may even provide some assistance in simplifying common series representations to their compact form.

The following problems exemplify the application of Picard's method. In both examples the problem is nonlinear.

EXAMPLE 3.12: NONLINEAR SPRING-DASHPOT SYSTEM

Consider a nonlinear square-law spring and dashpot system. The governing differential equation is

$$\frac{dx}{dt} = f(x, t) = B - Ax^2, \qquad x(0) = x_0 = 0. \tag{3.140}$$

Applying Picard's method, Eq. (3.139) allows the integration to produce an initial approximation function, x_1:

$$x_1 = x_0 + \int_0^t \left(B - Ax_0^2\right) dt = Bt. \tag{3.141}$$

Repeating this iteratively yields improvements in this function, or

$$x_2 = 0 + \int_0^t \left[B - Ax_1^2\right] dt = \int_0^t [B - A(Bt)^2] \, dt$$

$$= Bt - \frac{AB^2}{3} t^3,$$

$$x_3 = 0 + \int_0^t \left(B - Ax_2^2\right) dt = \int_0^t [B - A(Bt - AB^2 t^3/3)^2] \, dt$$

$$= \left[Bt - \frac{AB^2 t^3}{3} + \frac{2A^2 B^3 t^5}{15}\right] - \frac{A^3 B^4 t^7}{63}$$

$$\vdots$$

3.6. Picard's Method

The terms within [ ] are called stationary terms. The last term outside the braces may vary on the fourth approximation. Each subsequent integration generally adds to the number of stationary terms. The student may wish to compare this solution with that obtained in Section 3.2.2 [Eq. (3.66)].

EXAMPLE 3.13: BEAD ON A WIRE

We shall reexamine the problem of the bead on a wire. This problem is described by a second-order nonlinear differential equation,

$$\ddot{y} + ay + by^3 = 0. \tag{3.142}$$

We now transform (change) variables and obtain two first-order differential equations. We do this by introducing the velocity, $v = dy/dt = \dot{y}$, resulting in the new differential equations,

$$\frac{dy}{dt} = v, \qquad \frac{dv}{dt} = -ay - by^3.$$

For the special case of $a = 10$, $b = 100$, $y(0) = 1$, $v(0) = 0$, we obtain the differential equation

$$\frac{dv}{dt} = -10y - 100y^3. \tag{3.143}$$

The right-hand sides of these two equations are both analytic, allowing us to apply Picard's method. The first integration via Picard's method yields

$$y_1 = y_0 + \int_0^t v_0 \, dt = 1 + \int_0^t (0) \, dt = 1,$$

$$v_1 = v_0 - \int_0^t \left(10 y_0 + 100 y_0^3\right) dt = -\int_0^t (10 + 100) \, dt = -110t.$$

Repeating this procedure, we iteratively evaluate v_2, y_2, then v_3, y_3, then ... etc.:

$$y_2 = 1 - \int_0^t (110t) \, dt = 1 - 55t^2, \tag{3.144}$$

$$v_2 = -\int_0^t \left(10 y_1 + 100 y_1^3\right) dt = -110t. \tag{3.145}$$

Carrying this one more step, we obtain the approximation

$$y_3 = 1 - 55t^2 + \frac{0.8525}{6} t^4, \tag{3.146}$$

$$v_3 = -110t + \frac{17{,}050}{3} t^3 - 181{,}500 t^5 + \frac{16{,}637{,}500 t^7}{7}. \tag{3.147}$$

It is obvious that the application of Picard's method to higher-order systems is possible.

3.7 REVERSION OF POWER SERIES

In most direct solution techniques, there is no assurance that the final solution for the dependent variable, x, will be obtained as a functional in terms of the independent variable, t, or

$$x(t) = g(t). \tag{3.148}$$

Occasionally, the solution appears in the inverse form

$$t = f(x). \tag{3.149}$$

The technique by which this inversion of variables may undergo a reversion process to a proper form using power series assumes that the solution function is known but of the form $t = f(x)$, where the function f is generally a complicated and nonlinear expression. This function also is assumed to be expandable as a convergent power series. This is represented as

$$t = f(x) = A_0 + A_1(x-a) + A_2(x-a)^2 + \cdots + A_n(x-a)^n. \tag{3.150}$$

The goal is to find an inversion technique to map this to the more usual form of the solution,

$$x = \mathcal{G}(t) = \sum_0^N a_k(t - A_0)^k. \tag{3.151}$$

We assume that $A_1 \neq 0$ and seek to expand $(x-a)$ in a power series in $(t-A_0)$. The following intermediate variables are defined:

$$T = \frac{t - A_0}{A_1},$$
$$X = (x - a),$$
$$a_i = A_i/A_1, \quad i = 1, n.$$

The given series may be written as

$$T = X + a_2 X^2 + a_3 X^3 + \cdots. \tag{3.152}$$

The solution format, however, is desired in the form

$$X = b_1 T + b_2 T^2 + b_3 T^3 + \cdots. \tag{3.153}$$

Substitution of the above equation into the preceding one yields

$$T = (b_1 T + b_2 T^2 + b_3 T^3 + \cdots) + a_2(b_1 T + b_2 T^2 + b_3 T^3 + \cdots)^2$$
$$+ a_3(b_1 T + b_2 T^2 + b_3 T^3 + \cdots)^3 + \cdots + a_n(b_1 T + b_2 T^2 + b_3 T^3 + \cdots)^n.$$

3.8. Summary Comments

We now rearrange this into ascending powers of T and equate like terms on the left and right sides of the above expression. This yields

$$b_1 = 1,$$
$$b_2 = -a_2 = -\frac{A_2}{A_1},$$
$$b_3 = -2b_2 a_2 - a_3 = 2\left[\frac{A_2}{A_1}\right]^2 - \frac{A_3}{A_1},$$
$$b_4 = -5a_2^3 + 5a_2 a_3 - a_4 = -5\left[\frac{A_2}{A_1}\right]^3 + 5\frac{A_2 A_3}{A_1^2} - \frac{A_4}{A_1}.$$

The above coefficients are seen as applicable to the method irrespective of the original solution representation, $t = f(x)$. To demonstrate the method, the following example is presented.

EXAMPLE 3.14: SERIES REVERSION

Consider a solution function that is given in the following expression:

$$t = \exp(x) = 1 + x + x^2/2! + x^3/3! + \cdots. \tag{3.154}$$

The reversion method first requires us to define T:

$$t - 1 = T = x + x^2/2! + x^3/3! + \cdots. \tag{3.155}$$

In this case, the coefficients have the values $A_0 = 1$, $A_1 = 1, \ldots$. Therefore, $a_2 = 1/2$, $a_3 = 1/6$, $a_4 = 1/24, \ldots$. This results in

$$b_1 = 1,$$
$$b_2 = -a_2 = -1/2,$$
$$b_3 = -2a_2^2 - a_3 = 1/3,$$
$$b_4 = -5a_2^2 + 5a_2 a_3 - a_4 = -1/4, \ldots \text{etc.}$$

Finally, we write the solution as

$$x = (t-1) - \frac{(t-1)^2}{2!} + \frac{(t-1)^3}{3!} - \frac{(t-1)^4}{4!} + \cdots. \tag{3.156}$$

This is recognized as the power series of $\log(t)$. This solution function has a radius of convergence of $t < 2$.

3.8 SUMMARY COMMENTS

This chapter has focused on exact analytical solution techniques. Many students of nonlinear systems never consider the possibility that an analytic, exact solution is possible, and thus never make the effort to attempt one. Such solutions are invaluable, however, and the availability of computational algebra systems has made analytical solutions more possible.

This chapter also has presented examples demonstrating the unusual behavior of nonlinear systems. These were introduced in Chapter 2, Section 2.1, but the examples presented are intended to familiarize the reader with how the properties of nonlinear systems manifest themselves in a broad spectrum of different systems.

Throughout this chapter, the concept of normalization of differential equations introduced in Chapter 2, Section 2.6, was applied. In every case, irrespective of the form of solution sought, the method provided insight into the domain of the problems.

As a general practice, it is valuable to *search* for an exact solution, even if the problem or model must be grossly simplified to attain a soluble system of equations. The insight into the importance of the parameters that can be gained by studying the solution functions is generally well worth the effort. In some instances, an exact solution to a simplification of the problem provides all of the information sought. As was stated in Section 2.2, the math modeler must understand the purpose of the model, the resources available, and what information is most important. Within the boundaries of these constraints, an exact solution is always preferred if it can be attained.

3.9 PROBLEMS

3.1 In the system shown in Fig. 3.10, a mass, m, is located in the center of a tight uniform wire. The mass is constrained to move in a straight line. The system has the following properties:
- Cross-sectional area of wire $= A$,
- Modulus of elasticity of wire $= E$,
- Tension in wire when $y = 0$ is $= T$,
- Nominal length of wire at tension $T = L$.

Submit a solution with the following instructions:
(1) Derive the exact equation of motion for this system. Neglect gravitational effects to simplify the problem.
(2) Expand the nonlinear portion of the equation in a Taylor series about $y = 0$ up to and including the term in y^3.

3.2 Consider the growing phenomenon described in Problem 2.9.
(1) Solve the given equation to find x as a function of t.
(2) Using the computer, plot the results over the time period $t[0, 40]$ using the parameter values $A = 200$, $X_0 = 20, 100, 200, 300$, and $k = 0.002$.
(3) Discuss your results. Does this system behave in a manner you would or would not expect? Are there different regions in which the solution behaves differently? Do the parameters have physical meaning? Explain.

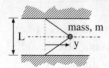

Figure 3.10. A mass suspended on a wire.

3.3 Consider the aircraft landing described in Problem 2.1.
   (1) Find the exact analytic solution $x(t)$ in terms of the problem variables. Do not insert the numerical values at this stage of the solution.
   (2) Linearize the problem and obtain an exact solution to the linear problem. Do not insert the numerical values at this stage of the solution.
   (3) Present both solutions graphically using the numerical values given.
   (4) Discuss your results and comment on the radius of convergence of the linearization.

3.4 Consider a mass resting on a cubic spring whose governing equation is

$$m\frac{d^2x}{dt^2} + bx^3 = 0, \qquad \frac{dx}{dt}(0) = 0, \qquad x(0) = A, \tag{3.157}$$

where m, b, and A are constants. Use a symbolic algebra language and solve this equation by the method of separation of variables, thereby obtaining an exact solution.

3.5 Using Picard's method, solve for the motion of a mass on a cubic spring as described in Problem 3.4. Include terms up to and including t^2.
   (1) Analytically compare these two solutions.
   (2) Using as data $m = 1$, $b = 1$, and $A = 5$, compare the exact and Picard's solutions for time $t = 0.5$ to 18.0.
   (3) Plot your results.

3.6 Consider the growing phenomenon described in Problem 2.9.
   (1) Transform the differential equation using scaling (normalization) transformations given by

$$\xi = x/X, \qquad T = t/\tau.$$

   (2) Identify the values of the scaling variables, X, τ, that reduce the differential equation and initial condition to the form

$$\frac{d\xi}{dT} - \xi + \xi^2 = 0, \qquad \xi(0) = \xi_0.$$

   Be certain to state the reasons for your selection of X, τ and also identify the analytic form of the initial condition, ξ_0.

   (3) Using a symbolic algebra program, obtain an exact solution to the differential equation and plot the results ξ vs T using the following data:

$$\xi_0 = 0.01, 0.3, 0.5, 1.0, 1.33,$$
$$T = T[0, 10].$$

   (4) Discuss your results. How has the normalization changed your perspective about this problem? Will your plots, in the form requested, change if A or k changes? Explain.

3.7 Consider the potential difference across the diode-capacitor circuit shown in Fig. 3.11.
   (1) Derive a math model of the circuit and show that it is in the form

$$C\frac{de}{dt} + ae + be^2 = 0, \qquad e(0) = E_0.$$

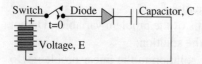

Figure 3.11. Simple switched diode-capacitor circuit.

(2) Normalize the differential equation and place it into the form
$$\frac{dy}{dx} + Dy^2 + y = 0, \qquad y(0) = 1.$$

3.8 The equation of motion for a metering pin damper is given as
$$\frac{d^2x}{dt^2} + \|x\|\frac{dx}{dt} + x = 0. \tag{3.158}$$

Find the exact solution to this problem. It can be written in the form
$$x = x_m \left[\frac{e^{ct} - 1}{e^{ct} + 1}\right], \tag{3.159}$$

where x_m and c are constants that you are to determine analytically.

3.9 The equation of motion for a mass supported on a cubic spring is described in Problem 3.4.
  (1) Eliminate time from the equation, and by use of the substitution $y = Y_0 \cos\phi$, find the integral expression for the period.
  (2) Set $y(0) = Y_0 = 1$ and $b = 100$. By the use of tables or using numerical integration, find $y(t)$ and plot it versus time for the first quarter of a cycle of oscillation.

3.10 The equation of motion for a mass supported on a damper and square-law spring may be written as
$$\frac{d^2y}{dt^2} + c\frac{dy}{dt} + ay + by^2 = 0,$$

where a, b, and c are constants and $y(0) = Y_0$, $dy(0)/dt = V_0 = 0$.
  (1) Solve this problem analytically using the power-series method.
  (2) Set $y(0) = Y_0 = 1$, $a = 10$, $b = 100$, and $c = 0$. Compare the results of your analysis to that obtained using the elliptic integral approach. Find $y(t)$ for these two cases and plot y/Y_0 vs time for the first quarter-cycle of oscillation.
  (3) Discuss your results.

3.11 Consider Problem 2.9 describing a growing phenomenon with finite resources. Solve this equation by means of Picard's method for terms up to and including t^2. Check the answer obtained in the following two ways:
  (1) Make a power-series expansion of the exact solution and compare it with the Picard's solution.
  (2) Using as data $k = 0.002$, $A = 200$, and $X_0 = 10$, compare the exact and Picard's solutions at time $t = 0.5$, 1.0, and 2.0.

3.12 Consider the problem of cooling of the Earth introduced in Problem 2.5.
  (1) Linearize the normalized model to obtain a simple differential equation that you can solve exactly.
  (2) Estimate the drop in the Earth's temperature over the next 100 years. Comment on the appropriateness of the linearization (Hint: What is the radius of convergence of the linearization?).
  (3) Estimate the time period over which the Earth will cool.

3.9. Problems

3.13 An equation to describe the cooling of the earth is

$$\frac{dT}{dt} = a^4 - T^4, \qquad T(0) = 1,$$

where a is a constant.
   (1) Solve this equation by means of Picard's method for terms up to and including t^2.
   (2) Obtain an exact solution. Make a power-series expansion of the exact solution and compare it with the Picard's solution.
   (3) Using $a = 10$, plot both exact and Picard's solutions over the domain $t(0, 2)$.

CHAPTER 4

Numerical Solution Methods

The solution of differential equations using numerical integration has become an accepted practice, almost to the extent that such solutions often are considered to be "exact solutions." It is for this reason that this material is included immediately following the chapter on exact solutions.

In light of the large body of literature on numerical methods for both linear and nonlinear systems, no attempt is made to cover this subject in great depth. Rather, the basic concepts upon which all modern numerical methods are based is presented along with examples intended to demonstrate a comparison of numerical methods with known exact and approximate solutions.

4.1 TAYLOR-SERIES METHOD

Taylor's method (see Section 3.1.1) is a general technique that allows one to compute a solution to accuracy levels that are limited by numerical considerations, not by attributes of the method proper. It is a method of great historical significance as well as serving as the basis upon which many other numerical integration methods are founded.

As a mental exercise, imagine that the solution of the differential equation

$$\frac{dx}{dt} = f(x, t) \tag{4.1}$$

is known and that this hypothetical solution is shown graphically in Fig. 4.1. If it is assumed that the solution is well known up to the point t_0, then the solution function can be expanded in a Taylor series about this point to find the solution at some slightly greater time, $t_0 + \delta$. This series can be written as

$$x(t_0 + \delta) = x(t_0) + \dot{x}(t_0)\frac{\delta}{1!} + \ddot{x}(t_0)\frac{\delta^2}{2!} + \cdots + \frac{d^n x}{dt^n}(t_0)\frac{\delta^n}{n!} + R_n. \tag{4.2}$$

This solution only converges to x at time $t_0 + \delta$ if the remainder, $R_n(t_0)$, goes to zero as $n \to \infty$, or

$$\lim_{n \to \infty} R_n(t) = 0. \tag{4.3}$$

In this instance, the solution function itself is found by the Taylor-series method:

$$x(t) = \sum_{m=0}^{\infty} \frac{1}{m!} \frac{d^m x(t)}{dt^m}. \tag{4.4}$$

The Taylor expansion represents an exact solution to the original problem. The accuracy of the method can be controlled by the number of terms that are used, and an estimate

4.1. Taylor-Series Method

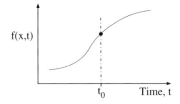

Figure 4.1. Solution to nonlinear system problem.

of the error is readily made by computing the magnitude of the residual (see the earlier Section 3.1.3). One also may compute the radius of convergence for the series approximation and find the resulting "solution" to have a limited range of usefulness.

The real limiting factor in the application of this method, however, lies in the requirement that the function $f(x, t)$ in Eq. (4.1) be a continuously differentiable function. In many real problems, it may not be impossible to obtain analytic derivatives. It is then common practice to numerically differentiate this function to obtain the higher-order derivatives. It is equally common for this method to suffer from the insertion of "noise" in Eq. (4.4) resulting from this numerical procedure. This situation may be improved by employing methods that do not require derivatives at all, a concept to be studied further in later sections and chapters.

EXAMPLE 4.1: NONLINEAR SPRING-DASHPOT MODEL

Consider the following differential equation:

$$\frac{dx}{dt} + Ax^2 = B, \qquad x(0) = 0. \tag{4.5}$$

For the sake of numerical simplicity, we arbitrarily set $A = B = 1$. The derivatives of x may be found as

$$\frac{dx}{dt} = 1 - x^2,$$

$$\frac{d^2 x}{dt^2} = 2x^3 - 2x,$$

$$\frac{d^3 x}{dt^3} = -2 + 8x^2 - 6x^4,$$

$$\vdots$$

etc.

Using the first m derivatives, the solution is

$$x(n\delta) = x[(n-1)\delta] + \sum_{k=1}^{m} \frac{d^k x}{dx^k} \frac{(n-1)\delta}{k!} \delta^k, \tag{4.6}$$

where δ is the step size and $x(n\delta)$ is the solution for $t = n\delta$. The solution is now assembled using just the first three derivatives. If the step size is chosen as $\delta = 0.1$, the following is obtained:

$$x(n\delta) = x[(n-1)\delta] + x'[(n-1)\delta]\delta + x''[(n-1)\delta]\delta^2/2! + x'''[(n-1)\delta]\delta^3/3!,$$
$$x(0.1) = 0 + 0.1 + 0 - 2/6000$$
$$= 0.099667,$$

TABLE 4.1. Numerical Solution of $dx/dt + x^2 = 1$, $x(0) = 0$

A. Taylor Series $x((n+1)\delta) = x(n\delta) + \frac{\delta^2}{2!}x''(n\delta) + \frac{\delta^3}{3!}x'''(n\delta)$

Time, t	Numerical approx., $x(t)$	$x'(t)$	$x''(t)$	$x'''(t)$	Exact soln. $\tanh(t)$	Error %
0.0	0.0000	1.0000	0.0000	−2.0000	0.0000	0.0000
0.1	0.0997	0.9900	−0.1974	−1.9405	0.0997	0.0000
0.2	0.1974	0.9610	−0.3794	−1.6972	0.1974	0.0000
0.3	0.2913	0.9151	−0.5331	−1.3642	0.2913	0.0000
0.4	0.3799	—	—	—	0.3800	0.0001

B. Modified Euler $x(n+1) = x(n-1) + 2x'(n)\delta$

Time, t	Numerical approx., $x(t)$	$x'(t)$	$x''(t)$	Exact soln. $\tanh(t)$	Error %	Comments
0	0.0000	1.0000	—	0.0000	0.0000	2-Term Taylor's
0.1	0.1000	0.9900	0.1980	0.0997	−0.0003	$x_1 = x_0 + x'(0)\delta$
0.2	0.1980	0.9608	0.1922	0.1974	−0.0006	
0.3	0.2922	0.9146	0.1829	0.2913	−0.0009	Modified Euler
0.4	0.3809	—	—	0.3800	−0.0009	

$$x(0.2) = x(0.1) + x'(0.1)/10 + x''(0.1)/200 + x'''(0.1)/6000$$
$$= 0.197366.$$

Refer to Table 4.1 for the remainder of the calculations and a comparison to the exact solution for this problem. The exact solution is

$$x(t) = \frac{e^{2t} - 1}{e^{2t} + 1} = \tanh(t). \tag{4.7}$$

Note that adding additional terms to the Taylor solution generally improves the numerical solution. One must be careful, however, because the errors associated with the computation of polynomials do not monotonically decrease with the order of the solution.

4.2 EULER METHOD

The general problem addressed by numerical methods is to find a solution to the differential equation

$$\frac{dx}{dt} = f(x) = \dot{x}, \tag{4.8}$$

where $x(t_0)$ is a known function. Euler's method is perhaps the earliest known technique for numerical integration of this differential equation. Leonhard Euler, perhaps the most famous mathematician in history, published almost 900 works in his time, exclusive of his correspondence and collaborative work with his son. In his textbook published in 1755, *Institutiones Calculi Differentialis*, Euler demonstrated the importance of the Taylor-series

4.2. Euler Method

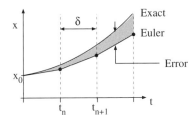

Figure 4.2. Cumulative numerical errors arising from Euler's method.

approach in the study and solution of differential equations. Euler's method employs a simplification of the Taylor-series approach, and begins with a series solution written as

$$x(t) = x(t_0) + \frac{\dot{x}(t_0)}{1!}\delta + \frac{\ddot{x}(t_0)}{2!}\delta^2 + \cdots, \quad (4.9)$$

where $\delta = t - t_0$. To avoid having to take higher-order derivatives and suffer the resulting numerical noise, the series is merely truncated after two terms. The resulting errors are hypothesized to be small, provided $\delta \ll 1$. In this approach, it is assumed that all of the eliminated terms are much smaller than the first term. This results in the following series of time-step estimates:

$$x(\delta) = x_1 = x_0 + \dot{x}(0)\delta,$$
$$x(2\delta) = x_2 = x_1 + \dot{x}(t_0 + \delta)\delta,$$
$$\vdots$$
$$x_{n+1} = x_n + \dot{x}(t_0 + n\delta)\delta.$$

The estimation of a new point at $(n + 1)$ steps therefore is always based on the past step, as depicted below:

$$x_{n+1} = x_n + \dot{x}(t_n)\delta. \quad (4.10)$$

The errors resulting from this numerical approximation are cumulative. This is shown diagrammatically in Fig. 4.2. This same effect holds for all numerical integration methods where an error in one step compounds, or adds to, the error in prediction of the subsequent steps. The error resulting from this approximation may be estimated by computing the magnitude of the first term dropped from the Taylor expansion:

$$\text{Error} = \ddot{x}(t_0)\delta^2/2!. \quad (4.11)$$

Note that this error is proportional to δ^2. Other methods may be employed to reduce this error, but they generally increase the computational overhead.

4.2.1 Modified Euler Method

In an attempt to improve on the accuracy of this method, it is hypothesized that by averaging the results of a forward application of Euler's method with a backwards step, the error will be reduced. The two steps (Euler's method backward and forward) are shown in Fig. 4.3 to be to $x(n - 1)$ and $x(n + 1)$, respectively. The results of these steps are written analytically as

$$x_{n-1} = x_n - \delta\dot{x}_n, \quad (4.12)$$
$$x_{n+1} = x_n + \delta\dot{x}_n. \quad (4.13)$$

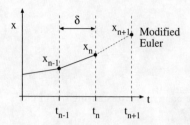

Figure 4.3. Modified Euler method applied using forward and backward differences.

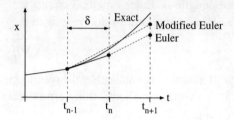

Figure 4.4. Comparison of Euler and modified Euler methods.

Subtracting the above two expressions, we obtain

$$x_{n+1} - x_{n-1} = 2\delta \dot{x}_n \tag{4.14}$$

or

$$x_{n+1} = x_{n-1} + 2\delta \dot{x}_n. \tag{4.15}$$

If the error is estimated as before, we find it to be proportional to δ^3. This reduces the δ^2 error of Euler's method, but a problem is introduced in starting the method since there is no way to infer knowledge about the solution at times less than t_0. The simpler Euler's method therefore is used to start the problem, and a switch is made to the modified Euler's method after two steps. The improved accuracy is depicted graphically in Fig. 4.4.

As shown in Table 4.1, the exact solution to a simple nonlinear problem is compared to that obtained using the Taylor-series method with three terms in Section 4.1 and the modified Euler method in Section 4.2.1. As with the Euler method, the modified Euler method must be started since it requires knowledge of the solution function at two previous steps. Note that the modified Euler method is started in this case by a two-term Taylor-series solution. Also note that the two-term error is larger than the three-term error for this initial time step.

4.2.2 Extension of Euler Method to Higher-Order Systems

Consider the system described by the two following coupled, first-order differential equations:

$$\frac{dy}{dt} = f_1(y, v), \tag{4.16}$$

$$\frac{dv}{dt} = f_2(y, v). \tag{4.17}$$

As a first-order approximation, we write

$$y(n+1) = y(n) + \dot{y}(n)\delta, \tag{4.18}$$

$$v(n+1) = v(n) + \dot{v}(n)\delta, \tag{4.19}$$

4.3. Runge-Kutta Method

TABLE 4.2. Application of Euler's Method to the Nonlinear Spring-Mass Problem

Time, t	Displacement, y	Velocity, v
0.00	1.000	0.000
0.01	1.000	−1.000
0.02	0.990	−2.000
0.03	0.970	−2.970
0.04	0.940	−3.883
0.05	0.901	−4.714

or

$$y(n+1) = y(n) + f_1(y_n, v_n)\delta, \tag{4.20}$$

$$v(n+1) = v(n) + f_2(y_n, v_n)\delta. \tag{4.21}$$

From these equations, it is possible to see that this method may be readily extended to higher-order systems. In the following example, we see this method applied to a second-order system.

EXAMPLE 4.2: NONLINEAR SPRING-MASS SYSTEM

Consider a mass on a spring having a stiffness proportional to the third power of the compression. The differential equation and initial conditions are given below:

$$\frac{d^2y}{dt^2} + 100y^3 = 0, \quad y(0) = 1, \quad \dot{y}(0) = v_0 = 0. \tag{4.22}$$

By defining $v = dy/dt$, the following are obtained:

$$\frac{dy}{dt} = v, \tag{4.23}$$

$$\frac{dv}{dt} = -100y^3. \tag{4.24}$$

Using (arbitrarily) $\delta t = 0.01$, the following numerical approximations result:

$$y_1 = y_0 + \delta \dot{y}_0 = 1 + (0.01)(0) = 1,$$

$$v_1 = v_0 + \delta \dot{v}_0 = 0 + (0.01)(-100) = -1.$$

The results of successive applications of Euler's method are shown in Table 4.2. Only a few time steps are given to demonstrate the method.

4.3 RUNGE-KUTTA METHOD

This method, attributed to both Runge[1] and Kutta,[2] is based on a weighted averaging technique for estimating the derivative function. Consider an arbitrary differential equation

---

[1] Carl David Tolmé Runge (1856–1927).
[2] Wilhelm Martin Kutta (1867–1944).

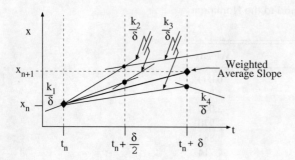

Figure 4.5. Graphical representation of the Runge-Kutta method.

and initial conditions represented symbolically as

$$\dot{x} = f(x), \qquad x(0) = x_0. \tag{4.25}$$

We fix δ and calculate the following:

$$k_1 = f[x(n)]\delta, \tag{4.26}$$
$$k_2 = f[x(n) + k_1/2]\delta, \tag{4.27}$$
$$k_3 = f[x(n) + k_2/2]\delta, \tag{4.28}$$
$$k_4 = f[x(n) + k_3]\delta. \tag{4.29}$$

Each of these k values represents a change in the dependent value over the interval based on derivatives at different positions within the whole step. It is possible to project a new value of the dependent variable using a weighted average of these estimated changes. Weighted averages are used to indicate the relative importance of each of these estimates of δx. The estimates within the middle of the step (k_2, k_3) are given the greatest weights (2) and the initial and final values (k_1, k_4) are assigned the smallest weights (1). This is represented by the equation

$$x(n+1) = x(n) + \left(\tfrac{1}{6}\right)(k_1 + 2k_2 + 2k_3 + k_4). \tag{4.30}$$

A pictorial representation of this procedure is shown in Fig. 4.5. It is generally accepted that the error of this method is of the order of magnitude of δ^5. To implement this algorithm, use the following recipe:

- Start at $x(n), t(n)$ and find the slope k_1/δ. Using this slope, go forward one-half step, $\delta/2$.
- At this point, find the slope k_2/δ and then go back to $x(n), t(n)$ and go forward one-half step, $\delta/2$.
- Find the slope k_3/δ at this point and, using it, go forward a full step (δ) from $x(n)$.
- Find the slope k_4/δ at this point.
- Average the four slopes found thus far, using weighting factors of 2 for the central slopes k_2/δ and k_3/δ.
- Using the average slope, compute from $x(n)$ to $x(n+1)$.

A simple example illustrates the application of this method.

4.3. Runge-Kutta Method

TABLE 4.3. Comparison of Exact Solution to Nonlinear Spring-Dashpot with Runge-Kutta Numerical Integration

t	Numerical approx., $x(t)$	k_1	k_2	k_3	k_4	Exact soln., $\tanh(t)$	Error %
0	0	0.1000	0.0998	0.0998	0.0990	0	0.00
0.1	0.0997	0.0990	0.0978	0.0978	0.0961	0.0997	0.00
0.2	0.1974	0.0961	0.0940	0.0940	0.0915	0.1974	0.00
0.3	0.2913	0.0915	0.0886	0.0887	0.0856	0.2913	0.00
0.4	0.3799	0.0856	0.0821	0.0823	0.0786	0.3800	0.03
0.5	0.4620	0.0786	0.0749	0.0751	0.0712	0.4621	0.02
0.6	0.5370	0.0712	0.0672	0.0674	0.0635	0.5371	0.02
0.7	0.6044	0.0635	0.0595	0.5980	0.0559	0.6044	0.00

EXAMPLE 4.3: NONLINEAR SPRING-DASHPOT PROBLEM

Consider the spring-dashpot model given in the following equation:

$$\frac{dx}{dt} = 1 - x^2, \quad x(0) = 0. \qquad (4.31)$$

The Runge-Kutta slope coefficients are given below and the results obtained are depicted in Table 4.3. With a time step of $\delta = 0.1$, the errors are of the order of magnitude of $\delta^5 = (0.1)^5 = 10^{-5}$. As can be seen in Table 4.3, the order of this estimate is confirmed by the analysis in this example:

$$k_1 = (1 - x_n^2)\delta,$$
$$k_2 = [1 - (x_n + k_1/2)^2]\delta,$$
$$k_3 = [1 - (x_n + k_2/2)^2]\delta,$$
$$k_4 = [1 - (x_n + k_3)^2]\delta,$$
$$x(n+1) = x(n) + (k_1 + 2k_2 + 2k_3 + k_4)/6.$$

EXAMPLE 4.4: TEMPERATURE-DEPENDENT RESISTANCE MODEL

This problem was initially derived in Example 2.7. The governing differential equations were transformed (normalized) to obtain

$$\theta^2 \frac{dI}{dy} = \frac{dE}{dy} I \left[2\theta \frac{d\theta}{dy} + 1 \right], \qquad (4.32)$$

$$\frac{d\theta}{dy} = R_\tau \left[R_{\text{pwr}} I^2 \theta^2 - \left(\theta - \frac{T_\infty}{T_0} \right) \right], \qquad (4.33)$$

where I is the dimensionless current, θ is the dimensionless temperature, and y is the dimensionless time. The coefficients R_τ and R_{pwr} are specific functions of the element values (resistance, capacitance, voltage) of the problem.

The results of a Runge-Kutta numerical analysis program were presented earlier in Section 2.6, Fig. 2.21. The specific values used in this example yield an electronic (current) response much shorter than the thermal (temperature) changes. Thus, the thermal response plays out over a time period of several time constants. Also note that most of the current change, approximately two-thirds, occurs within a period of one time constant. The initial choice for a numerical time step was $\delta = 0.01$. This choice was more than adequate for this solution.

The power of normalization as applied in the preceeding example thus is seen as a powerful adjunct to mathematical modeling. It permits us to evaluate the effects of different terms in the governing equations and to easily select a time step for the problem. In this example, there were really two timescales for the problem. The coefficients in one of the differential equations resulted in a ratio of these two time constants. Through this ratio, it is possible to more accurately select a time step for numerical solution of such a problem.

4.3.1 Solution of Two Simultaneous Equations

Extension of the Runge-Kutta method to a system of equations is straightforward, and we begin with two first-order differential equations:

$$\frac{dx}{dt} = f_1(x, y), \tag{4.34}$$

$$\frac{dy}{dt} = f_2(x, y). \tag{4.35}$$

These equations may be solved by computing the following:

$$K_1 = f_1[x(n), y(n)]\delta, \tag{4.36}$$
$$K_2 = f_1[x(n) + K_1/2, y(n) + L_1/2]\delta, \tag{4.37}$$
$$K_3 = f_1[x(n) + K_2/2, y(n) + L_2/2]\delta, \tag{4.38}$$
$$K_4 = f_1[x(n) + K_3, y(n) + L_3]\delta, \tag{4.39}$$
$$L_1 = f_2[x(n), y(n)]\delta, \tag{4.40}$$
$$L_2 = f_2[x(n) + K_1/2, y(n) + L_1/2]\delta, \tag{4.41}$$
$$L_3 = f_2[x(n) + K_2/2, y(n) + L_2/2]\delta, \tag{4.42}$$
$$L_4 = f_2[x(n) + K_3, y(n) + L_3]\delta, \tag{4.43}$$

where

$$x(n+1) = x(n) + (K_1 + 2K_2 + 2K_3 + K_4)/6, \tag{4.44}$$
$$y(n+1) = y(n) + (L_1 + 2L_2 + 2L_3 + L_4)/6. \tag{4.45}$$

4.3. Runge-Kutta Method

EXAMPLE 4.5: HARMONIC OSCILLATOR PROBLEM

Consider the governing differential equations for a harmonic oscillator:

$$\frac{dx}{dt} = -y, \tag{4.46}$$

$$\frac{dy}{dt} = -x, \tag{4.47}$$

with initial conditions $x(0) = 0$, $y(0) = 1$. The exact solution is $x = \sin(t)$, $y = \cos(t)$. By using the Runge-Kutta method, we compute the first time steps as:

$$x_1 = \delta - \delta^3/6 = \delta - \delta^3/3!, \tag{4.48}$$

$$y_1 = 1 - \delta^2/2 + \delta^4/24 = 1 - \delta^2/2! + \delta^4/4!. \tag{4.49}$$

4.3.2 Final Remarks on Runge-Kutta Method

Consider the arbitrary differential equation represented below:

$$\frac{dx}{dt} = f(x, t), \qquad x(0) = X_0. \tag{4.50}$$

The solution of differential equations of order >1 using Taylor expansions generally becomes an unstable, impractical affair if high-order derivatives are required. Such derivatives become quite complex, especially for nonlinear systems. Runge first introduced the approach that bears his name, and this work was extended by Kutta, who obtained specific weighting functions for the approximating functions for different orders of approximation. Gill derived another set of weights with the specific goal of reducing the memory requirements for the numerical solutions. The majority of the available computer programs available today still use Gill's constants, even though memory requirements generally are not serious limitations.

The general form of all Runge-Kutta methods is in the form

$$x_{i+1} = x_i + hW(x_i, t_i, h), \tag{4.51}$$

where W is termed the increment function and is simply a suitable approximation function over the interval. The usual approach is to assume some weighted sum of derivatives, as in

$$W = a_1 k_1 + a_2 k_2 + a_3 k_3 + a_4 k_4 + \cdots + a_n k_n = \sum_{i=1}^{N} a_i k_i. \tag{4.52}$$

Here, the weights a_i are constants and the derivatives k_i are chosen at suitable points over the interval. The approach used in the preceding section resulted in evaluations of four derivative functions at four different points within the boundary of a single step size: once at the beginning, t; twice at the half-step point, $t + \delta/2$; and once at the full step, $t + \delta$. The weights for the Runge-Kutta evaluations are $a_1 = a_4 = \frac{1}{6}$ and $a_2 = a_3 = \frac{1}{3}$.

4.4 MULTISTEP METHODS

The form of differential equation to be considered is

$$\frac{dx}{dt} = f(x,t), \qquad x(0) = X_0. \tag{4.53}$$

The repetitive application of Euler's method allows us to write the following:

$$x_1 = x_0 + \int_{t_0}^{t_1} f_0 \, dt, \tag{4.54}$$

$$x_2 = x_1 + \int_{t_1}^{t_2} f_1 \, dt, \tag{4.55}$$

$$\vdots \tag{4.56}$$

$$x_{i+1} = x_i + \int_{t(i)}^{t(i+1)} f_i \, dt. \tag{4.57}$$

If we have integrated for $k+1$ intervals, this also may be written as

$$x_{i+1} = x_{i-k} + \int_{t(i-k)}^{t(i+1)} G_i \, dt, \tag{4.58}$$

where $G(t)$ is the bar-graph function with ordinates

$$f_j = f(x_j, t_j), \qquad j = i-k, \ldots, i, \tag{4.59}$$

and the integral can be viewed as the area under the $G-t$ curve. If we replace the function $G(t)$ with an interpolating polynomial that passes through all of the points (x_j, f_j) for j near i, the area obtained should be very close to the original area. This interpolating polynomial is thus

$$L(t) = \sum_0^r a_i x^i \tag{4.60}$$

and

$$x_{i+1} = x_{i-k} + \int_{t(i-k)}^{t(i+1)} L(t) \, dt. \tag{4.61}$$

By defining a backward difference operator (∇) as follows

$$\nabla f(t) = f(t) - f(t-d),$$
$$\nabla^2 f(t) = \nabla f(t) - \nabla f(t-d),$$
$$\nabla^3 f(t) = \nabla^2 f(t) - \nabla^2 f(t-d),$$
$$\vdots$$
$$\text{etc.},$$

4.5. Step-Size Determination

and assuming that $t = t_i + a\delta$, and $dt = d(a\delta)$, we may use Newton's interpolation formula based on backward derivatives:

$$L(t_i + a\delta) = f_i + aVF_i + a(a+1)V^2 f_i/2! + a(a+1)(a+2)V^3 f_i/3! + \cdots$$
$$+ a(a+1)(a+2)\cdots(a+r-1)V^r f_i/r!.$$

This equation may be directly integrated and various values for k inserted ($k = 0, 1, 2, 3$) and for different orders of interpolating polynomial ($r = 1, 2, \ldots$), resulting in many equations for the projected value, $x(i+1)$, and an estimate of the error [the remainder, $(r+1)$ term]. Typical resultant equations thus may be derived as shown below for various values of r and k:

$$k = 0, \ r = 3: \quad x_{i+1} = x_i + (\delta/24)[55f_i - 59f_{i-1} + 37f_{i-2} - 9f_{i-3}] \quad (4.62)$$

with a remainder, R, on the order of (δ^3);

$$k = 1, \ r = 1: \quad x_{i+1} = x_{i-1} + 2\delta f_i \quad (4.63)$$

with a remainder term on the order of (δ^5). Note that this is also termed the modified Euler equation as described earlier in Section 4.2.1, Eq. (4.15):

$$k = 3, \ r = 3: \quad x_{i+1} = x_{i-3} + (4\delta/3)[2f_i - f_{i-1} + 2f_{i-2}] \quad (4.64)$$

with a remainder term on the order of (δ^5);

$$k = 5, \ r = 5: \quad x_{i+1} = x_{i-5} + (0.3\delta)[11f_i - 14f_{i-1} + 26f_{i-2} - 14f_{i-3} + 11f_{i-4}] \quad (4.65)$$

with a remainder term on the order of (δ^7).

All of these formulas use interpolating polynomials to integrate over the entire region through known points and for one step beyond. Thus, the method effectively extrapolates beyond the current step to the next. Note that they are not self-starting methods but must rely on some other technique to perform the first few integration evaluations. Once some historical information is determined, it is possible to switch to this open integration scheme. Notice the number of function evaluations that are required for each step.

4.5 STEP-SIZE DETERMINATION

The normalization of the describing equations has the beneficial property of making both the problem variables and their derivatives of order one. Because of this, the typical step size will be in the range $0.001 < \delta < 0.1$ unless the problem is essentially unstable or has been normalized improperly. An interesting question, however, is to ask what bounds one can place on the step size to yield an unconditionally stable solver.

Consider an ordinary differential equation (ODE) that will be used as a benchmark for comparing the different numerical integration schemes:

$$\dot{y} = \alpha y. \quad (4.66)$$

For Euler's method, we know that

$$\epsilon_{i+1} = \epsilon_i \left(1 + \delta \frac{\partial y}{\partial x}\right) + O(\delta^2),$$

where ϵ_i is the error at the ith step in x, and δ is the step size. For the numerical solution to be stable, we must have

$$\left|\frac{\epsilon_{i+1}}{\epsilon_i}\right| < 1.$$

For unconditional stability, this leads to $|1 + \alpha\delta| < 1$. Thus, for positive α, Euler's method is always unstable. This is because, without a corrector step, Euler's method always underestimates, as shown in Fig. 4.2. If $\alpha\delta$ is in a circle in the complex plane of radius 1, centered at $(-1, 0)$, then the algorithm is unconditionally stable, but not necessarily accurate. The solution can oscillate wildly about the true solution, but it will not diverge.

The parameter α has been plotted for many different algorithms, and can be of enormous value in circumscribing the region of stability of your algorithm, as applied to a specific problem. For instance, the implicit Euler method as well as the Adams-Moulton method (for $k = 1, 2$) are unconditionally stable for the benchmark ODE system of Eq. (4.66). Although this technique has only been studied intensively for a few ODE systems, it can be attempted when one is seeking a better understanding on the stability bounds for numerical step sizes.

4.6 SUMMARY COMMENTS

Although analytical solutions provide the greatest insight regarding the influence of various design parameters, one frequently must resort to numerical parametric studies to discern trends for parameters. The long and expensive trials (computation cycles, programming effort, time) that one must devise for this purpose often hide the very trends for which one is searching. In other instances, analytical solutions are impossible, and numerical solutions are the only avenue for understanding the behavior of such systems.

The practice of normalization of problems, introduced in Chapter 2 Section 2.6, also is found to be of enormous value when employing numerical solution techniques. Normalization provides important information into the domain of the problem, and frequently allows one to reduce the complexity because of the inherent understanding that arises about the importance of each term in the governing equations. But, perhaps the most important result of normalizing the formulation of a problem when using numerical methods is that the step size is no longer a total unknown.

4.7 PROBLEMS

4.1 Consider the cooling of the earth as described in Problem 2.5.
   (1) Using a numerical method, develop a solution to the normalized problem. Plot this over the dimensionless time, $t(0..2)$.
   (2) Compute the effective time constant of this problem. What is its significance?
   (3) Plot the drop in temperature (real units) over the next 100 years.

4.7. Problems

4.2 The equation of motion for a mass supported on a cubic spring may be written as

$$\frac{d^2y}{dt^2} + by^3 = 0,$$

where $b =$ constant and $y(0) = Y_0$, $dy/dt|_{y(0)} = V_0 = 0$.
(1) Reduce this second-order differential equation to two first-order coupled differential equations.
(2) Modify your numerical solution program from the preceding problem to be used as a general-purpose subroutine to solve two coupled first-order differential equations. You are to include a well-documented listing with your solution.
(3) Using this subroutine, solve this problem numerically and compare your results to those obtained using elliptic integrals in Problem 3.9. Use $b = 100$ and $Y_0 = 1$.

4.3 By the use of the Runge-Kutta method, solve the following simultaneous equations:

$$\frac{dx_1}{dt} = x_2, \quad x_1(0) = 1; \tag{4.67}$$

$$\frac{dx_2}{dt} = -0.1x_1 - x_1^3, \quad x_2(0) = 0. \tag{4.68}$$

(1) For time steps of 0.2, 0.5, and 1.0 solve for $x_1(t)$ and $x_2(t)$ for the time range $0 < t < 10$ s.
(2) Using the computer, plot your results two ways:
  (a) $x_1(t)$ and $x_2(t)$ versus t,
  (b) $x_2(t)$ vs $x_1(t)$ (ordinate vs abscissa).
(3) Discuss your results, explaining any differences in results between step sizes. Compare the amplitude and period with that for a linear system (drop the cubic x term).

4.4 Consider the problem described in Section 8.4.
(1) Write your own numerical integration scheme to analyze the variation of the level in the surge tank with time. Present your results graphically (plotted) using the computer. You may choose any of the numerical techniques introduced in class. So that you may use this routine again, it is suggested that you exercise care in writing the routine and that you write it as a subroutine in the high-level language of your choice. You are to electronically submit a listing of your program, a description of your findings, and your computer-generated plots.
(2) Take two extreme values for h': (a) its original steady value, $h_0 - (Q_{v0}/k)^2$, where Q_{v0} is the original steady flow rate; and (b) zero. Discuss your results in light of the physical problem and the nonlinearities involved.

4.5 Consider the following simultaneous linearized differential equations for the variation of dimensionless temperature, θ, and current, I, versus normalized time, y.

$$\frac{d\theta}{dy} = k(2I + 2\theta - 3), \quad \theta(0) = 1;$$

$$\frac{dI}{dy} = -2(k-1)\theta - (6k+1)I + 2(3k-1), \quad x_2(0) = 0.$$

TABLE 4.4. Property Data for the Resistor and Its Heat Transfer Coefficient

Variable	Value	Units
Heat transfer coefficient, h	10.0	W/s-cm^2-°C
Surface area, A	3.0	cm^2
Mass of resistor, m	10.0	g
Specific heat, C_p	0.3	W/g-°C
Nominal resistance, R_0	100.0	kΩ
Nominal temperature, T_0	23.0	°C
Capacitance, C	0.1	μF
Input voltage, E_i	10.0	V

For this analysis, the product of the electrical and thermal time constants and power factors is $k = R_{\text{pwr}} R_\tau = 2.0$.

(1) Using a symbolic algebra language, find the exact solution for $I(y)$ and $\theta(y)$.
(2) For time steps of 0.1, 0.5, and 1.0, numerically solve for θ and I over a time range of $0 < t < 5$ s.
(3) Using the computer, plot your results as θ and I versus y.
(4) Discuss your results, explaining any differences in results between step sizes, and comment on the exact solution.

4.6 Create a mathematical model for an RC circuit having a temperature-dependent resistance element described in Chapter 2, Example 2.7. A schematic of this circuit is shown in Fig. 2.19. Model the temperature-induced change in resistance using a power-law equation given as

$$R(T) = R_0 \left(\frac{T}{T_0}\right)^2,$$

where the resistance, R, is a function of the absolute temperature, T, and has the resistance value R_0 at a reference temperature T_0.

(1) Derive the governing differential equations for the temperature change and the current through the resistance. Neglect all current to the output, E_0.
(2) Normalize the governing differential equations using the following transformations:

$$\theta = \frac{T}{T_0}, \quad I = \frac{i}{i_0}, \quad y = \frac{t}{\tau}.$$

(3) Obtain the differential equations for dI/dy and $d\theta/dy$ as driven by the forcing function, E_I, or its derivative.
(4) Take the solution of this system as far as possible. Use a minimum of two solution techniques and produce a report of your findings.

Use the data from Table 4.4 for your calculations.

CHAPTER 5

Graphical Solution Methods

The use of graphical techniques has been the basis for much of the development in the fields of engineering and mathematics since their beginnings. Many of the earliest mathematical proofs known to exist were performed using graphically based word algorithms. The methods presented in this chapter are similarly founded upon purely graphical concepts which have produced some unique and interesting solution techniques. In view of the importance and wide use of interactive graphics, these methods are experiencing a resurgence of interest. All of the techniques presented are readily converted into interactive solution and display techniques. The value of visualizing solutions, especially over the entire solution domain, and the insight this provides cannot be overemphasized. The visual processing potential of the brain is brought into play when the information is presented in this form, and the solutions are more readily studied and integrated into a conceptual model of the system's behavior. The integration of computer graphics into the field of analysis therefore holds great promise for producing rapid solution techniques and providing a mechanism for visual interpretation of the character of the solution functions of nonlinear problems over broad domain fields.

The methods presented here are not completely general and are intended to be used for a class of first- and second-order nonlinear ordinary differential equations. This encompasses a large class of mechanical and dynamical systems, however, and is not as restrictive as it may at first appear. Indeed, if one introduces indicial mathematics and linear algebra (refer to Chapter 2, Section 2.2.8), one may see the potential for multidimensional extensions to the methods presented here. All of the methods presented are based on purely graphical and mathematical methods for estimating the solution trajectory from the slope of the trajectory. Like their numerical counterparts, the accuracy of the result is often dependent on the step sizes (arc lengths) used in the constructions.

5.1 FIRST-ORDER EQUATIONS (METHOD OF ISOCLINES)

We begin this chapter with a method that is the graphical equivalent of the Euler numerical method. As this and subsequent methods are developed, look for the parallels between the methods presented and the numerical methods presented earlier.

The system to be analyzed in this section is the rather general form of a first-order differential equation:

$$\frac{dx}{dt} = f(x, t), \qquad x(0) = X_0. \tag{5.1}$$

This certainly covers a broad class of problems since the function $f(x, t)$ may range from

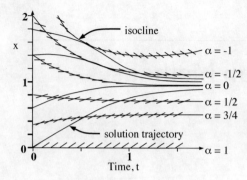

Figure 5.1. Method of isoclines.

simple to highly complex and from linear to nonlinear. The slope of the solution function curve at any point x_1, t_1 can be expressed as

$$\frac{dx}{dt}\bigg|_{x_1,t_1} = f(x_1, t_1). \tag{5.2}$$

As depicted in Fig. 5.1, the above mathematical expression can be used to graphically represent contour lines of constant slope, or *isoclines*. The name for these contours is derived from the Greek *iso*, meaning constant, and the verb *klinein*, meaning to lean. To further improve the visual interpretation of the isoclines, small line segments angled at the appropriate slope are plotted directly on the isoclines themselves. If the isoclines are selected appropriately within the solution domain, the full character of the solution over the solution plane is immediately obvious. Thus, the contour equations are given as

$$\frac{dx}{dt} = \alpha = f(x, t) = \text{constant} \tag{5.3}$$

and the lines they define are termed *isoclines*. The solution then may be derived by manually following the trajectory contours on the solution plane. The most interesting benefit of this method is the immediate understanding of the behavior of the system over the entire solution space.

The simplicity and value of this method is demonstrated in the following example.

EXAMPLE 5.1: NONLINEAR SPRING-DASHPOT PROBLEM (ISOCLINE)

Consider the system described by the differential equation and initial condition below:

$$\frac{dx}{dt} = 1 - x^2 = f(x, t) = \alpha, \qquad x(0) = 0. \tag{5.4}$$

The isoclines are thus contours represented by the solution to the equation

$$x = \sqrt{1 - \alpha}. \tag{5.5}$$

By substituting various numerical values for the slopes of the solution into the above equation, equations for lines representing the locus of points where the solution slope is a known constant are obtained. In this example, these lines are horizontal ($x = $ constant). Each time a solution trajectory crosses such an isocline, it must do so at the slope value

5.1. First-Order Equations (Method of Isoclines)

TABLE 5.1. Values of x for Isoclines of Slope α

Slope a	Soln. x
0	1.0
0.25	0.866
0.50	0.707
0.75	0.500
1.00	0.0
2.00	nonphysical

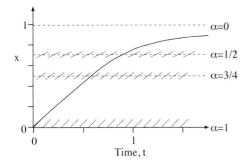

Figure 5.2. Isocline solution for a nonlinear spring-dashpot system.

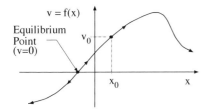

Figure 5.3. Phase-plane solution of $dx/dt = f(x)$.

of that isocline. These results are tabulated in Table 5.1 and depicted on the solution plane in Fig. 5.2.

5.1.1 Phase-Plane Analysis

A new solution technique also may be derived along similar lines using a somewhat reduced form of the differential equation,

$$\frac{dx}{dt} = f(x), \qquad x(0) = X_0, \tag{5.6}$$

into which the definition $v = dx/dt$ is interjected, resulting in the equation

$$v = f(x). \tag{5.7}$$

If a plot of the velocity function, v, as a function of the position, x, is an acceptable form of the solution, then this result may be plotted as shown in Fig. 5.3. A serious problem, however, is that the independent variable is lost from the analysis and no longer appears in any of the plots.

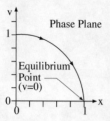

Figure 5.4. Solution of a massless nonlinear spring problem.

EXAMPLE 5.2: NONLINEAR SPRING-DASHPOT PROBLEM (PHASE PLANE)

Consider the problem of the previous example: a massless nonlinear spring with linear damping and a constant force. This is represented by the differential equation

$$\frac{dx}{dt} + x^2 = 1, \qquad x(0) = 0. \tag{5.8}$$

By defining a new variable (the velocity), $v = dx/dt$, this equation becomes

$$v = 1 - x^2. \tag{5.9}$$

To plot this in the phase plane, or the $v - x$ plane, one only needs the initial condition for the velocity, V_0, or position, X_0. These are related by

$$V_0 = 1 - X_0^2 = 1, \tag{5.10}$$

and the result is plotted in Fig. 5.4.

5.1.2 Recovery of the Independent Variable

As noted in Section 5.1.1, the loss of the independent variable results from the application of the phase-plane solution method. By invoking the relationship $v = dx/dt$, it is possible to separate the variables and obtain

$$dt = \frac{dx}{v}. \tag{5.11}$$

Integrating both sides yields

$$\int_{t_0}^{t_1} dt = \int_{x_0}^{x_1} \frac{dx}{v} = t_1 - t_0. \tag{5.12}$$

Thus, it is possible to plot $1/v$ versus x as shown in Fig. 5.5. The resulting area under the curve represents the independent variable for the solution.

EXAMPLE 5.3: NONLINEAR SPRING-DASHPOT PROBLEM (INDEPENDENT-VARIABLE RECOVERY)

Consider the preceding example of a massless spring. The resultant time, $t_1 - t_0$, is desired for the conditions $x(0) = X_0 = 0$ and $x_1 = 0.4$. By the method just introduced,

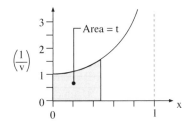

Figure 5.5. Phase-plane solution for the nonlinear spring-dashpot system.

this is seen to be obtained by

$$t_1 - t_0 = \int_0^{0.4} \frac{dx}{v} = 0.41. \tag{5.13}$$

Although this is an approximate method, the answer of 0.41 compares favorably with the exact answer which is $x = \tanh(t) = 0.42$.

5.2 SECOND-ORDER SYSTEMS

Consider the class of second-order differential equations given below:

$$\frac{d^2x}{dt^2} + f\left(x, \frac{dx}{dt}\right) = 0. \tag{5.14}$$

In general, these are termed autonomous types of equations. This implies that the independent variable does not appear explicitly in the differential equation.

5.2.1 Linear Spring-Mass System

Many second-order systems behave in a manner analogous to a common spring-mass mechanical system:

$$m\frac{d^2x}{dt^2} + kx = 0, \qquad x(0) = X_0; \qquad \frac{dx}{dt}(0) = V_0. \tag{5.15}$$

This may be rewritten as

$$x'' + (k/m)x = 0. \tag{5.16}$$

When compared to Eq. (5.14), it is seen that the resulting function is simply a linear function of x:

$$f\left(x, \frac{dx}{dt}\right) = \frac{k}{m}x. \tag{5.17}$$

As before, we set $v = dx/dt$, and this equation becomes

$$v\frac{dv}{dx} + \frac{k}{m}x = 0 \tag{5.18}$$

or

$$\frac{dv}{dx} + f(x, v)/v = 0. \tag{5.19}$$

This problem therefore may be solved in the phase plane!

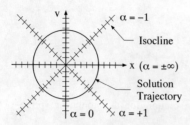

Figure 5.6. Isocline phase-plane solution for a harmonic system.

5.2.2 Method of Isoclines

We set $dv/dx = \text{constant} = \alpha = -f(x,v)/v$. This results in

$$\alpha v + f(x,v) = 0 \quad \text{(equation for isoclines)}. \tag{5.20}$$

Consider the problem given by the differential equation

$$f(x,v) = kx/m. \tag{5.21}$$

If we arbitrarily set $k/m = 1$, this results in the following equation for the isoclines:

$$\alpha v + x = 0 \longrightarrow v = -(1/v)x. \tag{5.22}$$

This produces isoclines that are simple radial lines through the origin. Each of these has a slope that is the inverse reciprocal of the solution slope, α. The solution trajectories are therefore always perpendicular to the isoclines themselves, producing a set of concentric circles, as shown on the phase plane in Fig. 5.6. The independent variable may be recovered using the relation

$$\int_0^t dt = \int_{X_0}^x \frac{dx}{v}. \tag{5.23}$$

The original problem, however, was

$$\frac{d^2x}{dt^2} + \frac{k}{m}x = 0. \tag{5.24}$$

In the general case, the parameters characterizing the system do not yield the special case considered above, that is, $k/m = 1$. Following the method outlined, the general isocline equation is found to be

$$\alpha v + (k/m)x = 0. \tag{5.25}$$

The phase-plane solution is therefore as shown in Fig. 5.7.

5.2.3 Effect of Normalization

In the preceding sections, it has been shown that

$$\frac{d^2x}{dt^2} + x = 0 \longrightarrow \text{Circles in phase plane} \tag{5.26}$$

5.3. Liénard's Method

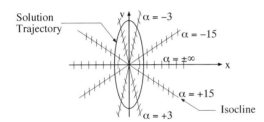

Figure 5.7. Nonnormalized phase-plane solution.

and

$$\frac{d^2x}{dt^2} + \frac{k}{m}x = 0 \longrightarrow \text{Ellipses in phase plane.} \tag{5.27}$$

Let us consider the independent-variable change to $T = \omega_0 t$, where ω_0 is a constant. This results in

$$\frac{dx}{dt} = v = \frac{dx}{d(T/\omega_0)} = \omega_0 \frac{dx}{dT} = \omega_0 z, \tag{5.28}$$

where $dx/dT = z$. Performing a similar operation for the second derivative results in

$$\frac{d^2x}{dt^2} = \omega_0^2 \frac{dz}{dT} \tag{5.29}$$

or

$$\omega_0^2 \frac{dz}{dx} + \frac{k}{m}x = 0. \tag{5.30}$$

This equation may be simplified if we arbitrarily select ω_0 such that $\omega_0 = \sqrt{k/m} \neq 0$ to obtain

$$\frac{d^2x}{dT^2} + x = 0. \tag{5.31}$$

This simple transform therefore reduced our solution to one having circles on the phase plane, as depicted in Fig. 5.6. The solution was originally obtained in Section 5.2.2 for the special case of $k/m = 1$, which produced a normalized differential equation. Contrast this with the result portrayed in Fig. 5.7 prior to normalization.

5.3 LIÉNARD'S METHOD

Many advances in mathematics have resulted from the intuition and insight of early mathematicians. In this instance, Liénard[1] was seeking solutions to a class of differential equations of the form

$$\frac{d^2x}{dt^2} + f_1\left(\frac{dx}{dt}\right) + x = 0. \tag{5.32}$$

---

[1] Alfred Marie Liénard (1869–1958).

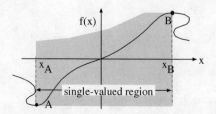

Figure 5.8. Example of a single-valued function.

He defined a velocity function, $v = dx/dt$, and obtained

$$\frac{dv}{dt} + f_1(v) + x = 0. \tag{5.33}$$

This may be rewritten in terms of the derivative

$$\frac{dv}{dx} = -\frac{[x + f_1(v)]}{v}. \tag{5.34}$$

The solution from this point on is based on the assumption that the function, f_1, from the differential equation is known and well behaved. The function is assumed to be single valued as shown in Fig. 5.8. Single-valued functions are those that only have a single result for any given value of the independent variable. In Fig. 5.8, the function is therefore only strictly single valued in the region $A - B$. The solution is derived from a simple observation based on a graphical interpretation. Consider the abstract representation of a nonlinear form of the function as shown in Fig. 5.9. To understand this method, consider the following recipe for the solution using Liénard's method:

1. Plot the function $f_1(v)$ rotated 90° on the phase plane.
2. From the initial condition point P_0, draw a line parallel to the x axis until it intersects the function $x = -f_1(v)$ at point c.
3. From c, draw a line parallel to the v axis until it hits the axis at point d.
4. Using d as the center, draw an arc of a circle from P_0 to some point P_1. This is a small piece of the solution trajectory.
5. Repeat this method from P_1.

Proof: The slope of the line $\overline{dP_0}$ is $V_0/[f_1(V_0) + X_0]$. The slope of the solution trajectory is

$$\left.\frac{dv}{dx}\right|_{X_0,V_0} = -\frac{X_0 + f_1(V_0)}{V_0}. \tag{5.35}$$

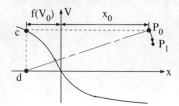

Figure 5.9. Function $x + f_1(v) = 0$ in the phase plane.

5.3. Liénard's Method

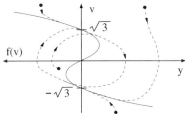

Figure 5.10. Solution of Van der Pol oscillator.

These two slopes are negative reciprocals; thus, they are always perpendicular. The solution then may be drawn as an arc using d as the center. It is upon this simple observation that the necessary and sufficient condition is found to graphically obtain the solution. This method allows for the solution of a very wide class of problems on the basis of a clever graphical technique.

Note that the direction of the solution is inferred from the values of x, v. If v is positive, then x must be increasing with time, as noted in Fig. 5.9.

EXAMPLE 5.4: VAN DER POL OSCILLATOR

Consider a van der Pol oscillator [51]. This is a nonlinear system whose governing differential equation is of the form

$$\frac{d^2 y}{dt^2} - \mu \left[\frac{dy}{dt} - \frac{1}{3} \left\{ \frac{dy}{dt} \right\}^3 \right] + y = 0. \tag{5.36}$$

If we set $v = dy/dt$, this reduces to

$$v \frac{dv}{dy} - \mu \left[v - \frac{v^3}{3} \right] + y = 0. \tag{5.37}$$

Thus, $f(v) = -\mu [v - v^3/3]$. The solution is depicted graphically in Fig. 5.10. The reader is encouraged to attempt the solution to this problem.

EXAMPLE 5.5: AUTOMOBILE BRAKING PROBLEM

Another example of the application of Liénard's method can be found by modeling the behavior of a simple brake system. In this problem, a brake pad is held against a rotating machine member. The member holding the brake pad has a nonlinear stiffness that gives rise to a nonlinear frictional force. An overview of the system is shown in the Fig. 5.11.

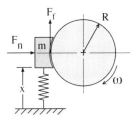

Figure 5.11. Simple brake system.

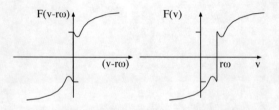

Figure 5.12. Typical frictional forces in a simple brake.

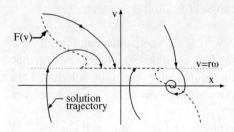

Figure 5.13. Liénard's solution for a simple braking problem.

The governing differential equation for this system is very similar to that of the familiar spring-mass-dashpot system:

$$\frac{dx}{dt} - v = 0, \tag{5.38}$$

$$m\frac{dv}{dt} + f[v - r\omega] + kx = 0. \tag{5.39}$$

The frictional force, F, is a highly nonlinear function as depicted in the Fig. 5.12. The frictional force shown in Fig. 5.12 results in the solutions depicted in Fig. 5.13 using the method of Liénard. The nonlinearity of the frictional force and the differential equation would pose serious problems for many other solution techniques. Note also that the solution technique has the added benefit of providing insight on the complete solution for any initial conditions in the phase plane. It also shows the existence of any limit cycles where the system reaches a stable cyclic behavior pattern.

5.4 PELL'S METHOD

This technique, introduced by Pell,[2] is used to solve problems concerned with the following type of second-order nonlinear ordinary differential equations:

$$\frac{d^2x}{dt^2} + f_1\left(\frac{dx}{dt}\right) + f_0(x) = 0, \tag{5.40}$$

or, in a slightly different form known as Pell's equation,

$$v\frac{dv}{dx} + f_1(v) + f_0(x) = 0. \tag{5.41}$$

This may be rewritten to obtain

$$\frac{dv}{dx} = -\frac{[f_0(x) + f_1(v)]}{v}. \tag{5.42}$$

---

[2] John Pell (1611–1685).

5.4. Pell's Method

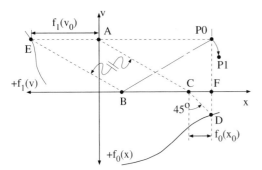

Figure 5.14. Pell's method.

This system may be solved graphically using the following "recipe":

1. Using $v = dx/dt$, plot $x = -f_1(v)$ and $v = -f_0(x)$ on the phase plane. Note that the slope of $\overline{BP_0}$ is $v_0/[f_1(v_0) + f_0(x_0)]$. The derivative is

$$\frac{dv}{dx} = -\frac{[f_0(x_0) + f_1(v_0)]}{v_0}\bigg|_{x_0, v_0}, \qquad (5.43)$$

and the slope of the solution trajectory is therefore perpendicular to the arc $\overline{BP_0}$.
2. Using B as the center, draw the arc $\overline{P_0 P_1}$.

Note that this solution technique is very similar to that of Liénard. In both methods, a solution technique was based on finding a curve form that presented a radius of curvature for the solution. This procedure is depicted graphically in Fig. 5.14.

EXAMPLE 5.6: SIMPLE PENDULUM PROBLEM

Consider the differential equation for a simple pendulum. This nonlinear equation was previously derived and is

$$\frac{d^2\theta}{dt^2} + \frac{g}{l}\sin\theta = 0. \qquad (5.44)$$

As before, the time is scaled by setting $\tau = \omega_0 t$, obtaining

$$\omega_0^2 \frac{d^2\theta}{d\tau^2} + \frac{g}{l}\sin\theta = 0. \qquad (5.45)$$

By a judicious choice of ω_0, one obtains the transformed differential equation for the pendulum as given by

$$\frac{d^2\theta}{d\tau^2} + \sin\theta = 0. \qquad (5.46)$$

This is reduced to a set of first-order equations by the transformation $V = d\theta/d\tau$, resulting in

$$\frac{dV}{d\theta} = -\frac{\sin\theta}{V} = \text{isocline slope} = \alpha \qquad (5.47)$$

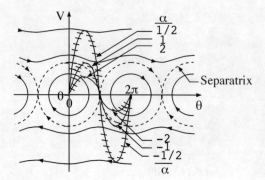

Figure 5.15. Phase-plane solution for a simple pendulum.

or

$$V = -\frac{\sin(\theta)}{\alpha}. \tag{5.48}$$

These isoclines are plotted in Fig. 5.15 and are seen to be simple harmonic functions. As before, the phase-plane solution does not produce information about the independent variable.

5.5 SUMMARY COMMENTS

In this chapter, various graphical methods are presented that represent graphical analogues to numerical methods. There exists a large spectrum of nonlinear problems that fall within the scope of the general analysis methods presented here. These methods offer an attractive, graphically based alternative to the usual Taylor or Euler numerical solution techniques. The advantages of the methods presented here are unique because of the form in which the results are presented. They are powerful, simple, visually based methods that provide remarkable domainwide insight into the behavior of the system under analysis. As with other methods of analysis, it is frequently more important to get an overview of the behavior of a system than to predict its response with great precision. No method other than the graphical methods presented here provides the immediate overview of the system behavior.

What is missing from these methods, as with their numerical competitors, is an analytical basis for optimizing the design of the system under analysis. Unlike numerical methods, however, the graphical nature of the methods frequently yields insights into the reasons why systems behave in the ways they do. This insight may be the expert knowledge needed to heuristically optimize system performance.

5.6 PROBLEMS

5.1 The equation for a Robinson's oscillator is

$$\ddot{x} + \left(\dot{x} - \frac{|\dot{x}|}{\dot{x}}\right) + x = 0.$$

Plot the response of this oscillator in the $(\dot{x}, x)$ phase plane.

5.2 The equation of motion for a simple nonlinear braking system may be written as

$$m\frac{dv}{dt} + F(v - r\omega) + kx = 0, \tag{5.49}$$

5.6. Problems

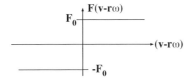

Figure 5.16. Bistate force generated by a frictional element acting on a rotating member with relative velocity $(v - r\omega)$.

where x = deflection of the brake-frame member; $v = dx/dt$, velocity of frictional brake element and frame; m = mass of frictional elements and frame (constant); F = force between brake element and frame; F_0 = maximum value of f; k = frame stiffness (constant); r = radius of rotating member (constant); ω = rotational speed (constant); t = independent variable, time.

Consider the special case where $m = 0.1$ slug, $k = 1000$ lbf/ft, $r\omega = 3$ ft/s, and a bistable frictional force exists in the form shown in Fig. 5.16 with a maximum force amplitude of $F_0 = 20$ lbf. Put the system equation into the Lienard form and solve it in the phase plane for several initial conditions. Submit a computer-generated plot of your results. You are also required to write a general-purpose routine that calls a user-defined function for $f(v)$ and solves the differential equation defining the Lienard problem,

$$\frac{d^2x}{dt^2} + f(v) + x = 0. \tag{5.50}$$

This system is subject to the initial conditions $x(0) = x_0$, $v(0) = v_0$. Your routines should have the following arguments:

BrakeFn$(F_0, v - r\omega) \to f$,

Lienard$(x, v - r\omega, \theta) \to v$,

where θ is the arc length of the Lienard solution from $x = x_0$, $v = v_0$ to the point x, v. Your Lienard routine is to call the BrakeFn routine and Lienard must be called iteratively to obtain $v(x)$. The results are to be plotted using a standard plotting package.

For additional credit, you might wish to compare this solution technique with solutions found using other techniques.

5.3 **Growing phenomena.** Solve Problem 2.9 using the method of isoclines for the specific case of $k = 0.01$ and $A = 300$. Use a variety of initial conditions, including $X_0 = 0, 20, 50, 100, 300, 400$. Extend your solution in the independent variable to cover the range $0 \le t \le 20$.

Write a program to draw the isoclines over the solution plane (x vs t) extending from $x(0, 300)$ and $t(0, 25)$ for the slopes

$$\frac{dx}{dt} = [-40, -30, -20, -10, -5, -1, 0, 1, 5, 10, 20].$$

Caution: Beware of both roots of SQRT( )!

5.4 **Bang-bang ship rudder control.** The equation of motion for a ship as depicted in Fig. 5.17 may be written as

$$I\frac{d^2\phi}{dt^2} + H\frac{d\phi}{dt} = M,$$

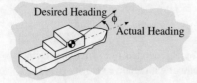

Figure 5.17. Bang-bang servo control of a ship's heading.

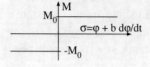

Figure 5.18. Bistate switching moment generated by a bang-bang servomechanism controlling a ship's rudder.

where ϕ = angle between actual and desired heading of ship; I, H = constants; M = moment due to rudder; t = time. Consider a pseudo bang-bang rudder control system with linear switching such that the moment, M, is a function of $\sigma = \phi + b(d\phi/dt)$ as shown in Fig. 5.18. This may be described as

$$M = M_0 f(\sigma)$$
$$= M_0 f\left(\phi + b\frac{d\phi}{dt}\right),$$

where

$$f = \begin{cases} +1 & \sigma < 0 \\ -1 & \sigma > 0. \end{cases}$$

(1) Put the equation in the form

$$\frac{d^2 X}{dT^2} + \frac{dX}{dT} = f\left[X + \frac{bH}{I}\frac{dX}{dT}\right].$$

(2) By the use of isoclines, plot the phase-plane solution of the equation in part 1 for the case in which $bH/I = 1/2$. Carefully examine the behavior of the system along the switching line $(X + \frac{1}{2}dX/dT) = 0$. Use a graphical plotting library or program for preparing your plots and depict your results over the domain $X[-4, 4], dX/dT[-3, 3]$. Include solution trajectories for the initial conditions for $X, dX/dT$ shown in Table 5.2.

TABLE 5.2. Initial-Condition Values for the Phase-Plane Solution

X	$\frac{dX}{dT}$
3.5	−1.0
3.5	−2.0
1.5	3.0
0.5	−2.0
−2.0	−2.0
−3.5	2.0

5.6. Problems

5.5 A boat weighing 3500 lb is traveling at $v = 15$ ft/s when its engine is stopped. The water resistance versus velocity is given in the table below. Plot a curve of velocity versus time, and solve for the time required for the boat to come to a stop.

velocity v, ft/s	0	2	4	6	8	10	12	14
resisting force f, lbf	18	19	22	28.2	48.3	113	280	497

5.6 A 200-lb car model is to be towed along a track using a cable wrapped around a 10-in.-diam drum, and directly driven by a d.c. motor. The vehicle is initially at rest, and the cable taut. The rolling resistance is almost constant at 15 lbf, and the torque of the motor as a function of speed is given in the table below. Find the velocity of the car as a function of time after the motor is started and the distance required to reach 90% of the final speed.

Drum speed, rpm	0	200	400	600	800	1000	1100	1150	1200	1250
Torque, ft-lbf	242	228	206	182	150	108	76	53.8	39.6	0

5.7 Consider the spring-mass system described in Problem 6.1, for $a = 0, b = 1$. For the case where $A > 0$, determine the range of stability of the system represented by this equation using two methods: (1) Plot the solution in the phase plane, and (2) use the method of isoclines to obtain a map of the solution plane.

5.8 The Rayleigh equation (obtained from the van der Pol equation) is

$$\ddot{y} - \mu\left(\dot{y} - \frac{\dot{y}^3}{3}\right) + y = 0. \tag{5.51}$$

By use of Lienard's method, find phase solution trajectories for several initial conditions for the special case of $\mu = 1$.

CHAPTER 6

Approximate Solution Methods

Most modelers of nonlinear systems do not consider seeking an analytic solution as a cost-effective approach. Consequently, the most common form of solution for a nonlinear system is a numerical one, and the solution generally must be repeated for each initial value or boundary condition as well as for any change in parameter. Thus, the numerical solution does not have general validity but is specific to the conditions for which it was obtained. The value of an analytic solution function, however, is that it permits one to perform parametric studies, vary the initial conditions, or perform a variety of *analytical operations* such as further integration, or differentiation.

The modern pupil of such methods has a new tool in his arsenal that previous generations have not had: The availability of computerized symbolic algebra and calculus programs allow one to avoid the long tedious operations of the past and obtain analytic solutions quickly and error-free. As a precursor to the use of an analytic approach, it is imperative that all pertinent relationships be reduced to an algebraic form in the formulation of the problem. The resulting solutions then may be expressed in terms of the parameters of this formulation.

In the following sections, various methods of solution are presented. They are all approximation methods and their accuracy is determined by the validity and scope of the approximations. As has been expressed previously, the value of direct analytic solutions is in their ability to provide insight into the behavior of the system under study. In this chapter, the methods presented are intended to lead to analytic solutions, but at the cost of making approximation assumptions. As in previous chapters, the examples chosen were selected to demonstrate the essential properties of nonlinear systems. Return to Chapter 2, Section 2.1, and study these properties before continuing so that you may look for them in the examples presented here.

In many of the examples, a symbolic calculus program was used. These computer tools remove much of the tedium of manual integration and algebraic manipulation. They still require, however, that the user understand the underlying mathematics and operations.

6.1 METHOD OF PERTURBATION

The method of perturbation is based on the concept that a zeroth-order solution may exist that represents the general behavior of the system under analysis. By subtraction of this behavior from the system's governing equations, one may obtain new solutions for incremental improvements to this fundamental behavior. This overall approach is termed a perturbation analysis. In the following sections, the systems in which such methods are useful are defined, the details of the method are introduced, and multiple examples are presented.

6.1. Method of Perturbation

6.1.1 Intended Area of Applicability

The method of perturbation assumes that the greatest portion of the response of the system under study may be represented by a solution to a reduced, linear form of the full nonlinear governing differential equation. The method thus is intended for almost-linear systems. One such differential equation is

$$\frac{dx}{dt} + ax + bx^2 = 0, \tag{6.1}$$

subject to the constraint that

$$\left|\frac{bx^2}{ax}\right| = \left|\frac{bx}{a}\right| \ll 1 \tag{6.2}$$

or

$$x \ll \frac{a}{b}. \tag{6.3}$$

6.1.2 Solution Technique

The solution technique for this method is straightforward. The initial step is to assume a power-series solution in the coefficient of the nonlinear term in the differential equation. For the preceding example given in Eq. (6.1), this solution is written as

$$x(t) = x_0(t) + bx_1(t) + b^2 x_2(t) + b^3 x_3(t) + \cdots. \tag{6.4}$$

This solution function is inserted back into the differential equation (6.1) and separated into terms of like powers of the nonlinear coefficient, b. Each of these equations then is solved in succession.

The technique may be demonstrated through the use of an example.

EXAMPLE 6.1: AUTOMOBILE HANDLING PROBLEM

The governing differential equation for the handling of an automobile (see the development in Chapter 2, Section 2.5.8) is written as

$$\frac{d\alpha_f}{dt} + A_1 \alpha_f + A_2 \alpha_f^2 = 0, \qquad \alpha_f(0) = \alpha_{f_0}. \tag{6.5}$$

This equation is of the same form as Eq. (6.1), or

$$\frac{dx}{dt} + ax + bx^2 = 0, \qquad x(0) = X_0. \tag{6.6}$$

In this example, the parameter b is a small term numerically, and thus, for $ax \gg bx^2$, it is possible to apply the method of perturbation. The form of the assumed solution is thus

$$x(t) = x_0(t) + bx_1(t) + b^2 x_2(t) + \text{HOT}. \tag{6.7}$$

In this equation, the x_0 term is the baseline or zeroth-order solution and x_1 is a first-order correction term, and so forth. The remaining terms are higher-order correction terms (HOT) and, for simplicity, the solution function is truncated after only three terms. Inserting this function into the differential equation yields

$$\left[\frac{dx_0}{dt} + b\frac{dx_1}{dt} + b^2\frac{dx_2}{dt} + \cdots\right] + a[x_0 + bx_1 + b^2x_2 + \cdots]$$
$$+ b[x_0 + bx_1 + b^2x_2 + \cdots]^2 = 0,$$

or, rearranging, one may obtain

$$b^0\left[\frac{dx_0}{dt} + ax_0\right] + b^1\left[\frac{dx_1}{dt} + ax_1 + x_0^2\right] + b^2\left[\frac{dx_2}{dt} + ax_2 + 2x_0x_1\right]$$
$$+ b^3[\cdots] + \cdots = 0. \tag{6.8}$$

A solution to this system is sought as $b \to 0$. Thus, all of the bracketed terms may be solved individually as indicated below:

$$dx_0/dt + ax_0 = 0 \to x_0 = ce^{-at},$$
$$dx_1/dt + ax_1 = -x_0^2,$$
$$dx_2/dt + ax_2 = -2x_0x_1,$$
$$\vdots$$

etc.

The issue of how to treat the initial conditions for each of the above solutions now must be developed. The original assumed form of the solution function, $x(t)$, and its initial condition are

$$x(t) = x_0 + bx_1 + b^2x_2, \qquad x(0) = X_0. \tag{6.9}$$

This must be valid for all b, even for $b = 0$. Thus, the zeroth-order solution is used to satisfy the initial conditions, and all subsequent correction terms are given homogeneous initial conditions. Thus,

$$x_0 = X_0 e^{-at}. \tag{6.10}$$

It is now possible to solve for each additional correction term:

$$x_1 = a\left[\frac{X_0}{a}\right]^2 [e^{-2at} - e^{-at}], \qquad \text{where } x_1(0) = 0,$$
$$x_2 = a\left[\frac{X_0}{a}\right]^3 [e^{-3at} - 2e^{-2at} + e^{-at}], \qquad \text{where } x_2(0) = 0.$$

Reconstructing the full system solution, $x(t)$, and gathering terms, we may write the result of the perturbation analysis as

$$x(t) = X_0 e^{-at}\left\{1 + \left[\frac{bX_0}{a}\right][e^{-at} - 1] + \left[\frac{bX_0}{a}\right]^2 [e^{-at} - 1]^2\right\}. \tag{6.11}$$

6.1. Method of Perturbation

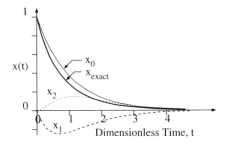

Figure 6.1. Perturbation solution to the automobile handling problem.

In this example, if one were to have used more correction terms, these added solution functions would develop in forms similar to the first three terms shown in Eq. (6.11). The radius of convergence of the solution may be defined by studying the ratio

$$\frac{bX_0}{a} < 1, \tag{6.12}$$

or, in another form,

$$bX_0^2 < aX_0. \tag{6.13}$$

This is exactly analogous to the original assumption made in this solution. The exact solution to the original problem may be shown to be

$$\alpha_f(t) = \frac{A_1 \alpha_{f_0}}{A_1 e^{A_1 t} + A_2(\alpha_{f_0} e^{A_1 t} - 1)}. \tag{6.14}$$

If this is expanded using the binomial theorem, the same solution function is derived as was found using the method of perturbation, Eq. (6.11). The exact solution and the various terms of the perturbation solution are depicted graphically in Fig. 6.1 for $X_0 = 1$, $A_2/A_1 = 0.5$.

In Example 6.1, Fig. 6.1 demonstrates graphically the features of the various terms that result from applying the perturbation method. In Example 6.2, more new information will similarly evolve but will pertain to the frequencies present in the response of a nonlinear system.

EXAMPLE 6.2: BEAD ON A WIRE

The following differential equation yields elliptic integrals as exact solutions:

$$\frac{d^2 x}{dt^2} + ax + bx^3 = 0, \quad x(0) = A, \quad \frac{dx}{dt} = 0. \tag{6.15}$$

It is assumed that the influence of bx^3 is less than that of ax, and a frequency term is defined as $\omega^2 = a$. The form of the solution also is assumed as

$$x(t) = x_0(t) + bx_1(t) + [b^2 x_2(t) + \cdots]. \tag{6.16}$$

This series is truncated beginning at the opening square bracket to obtain a finite series. Substitution of the above into the differential equation yields

$$b^0 \left[\frac{d^2 x_0}{dt^2} + \omega^2 x_0 \right] + b^1 \left[\frac{d^2 x_1}{dt^2} + \omega^2 x_1 + x_0^3 \right] = 0. \tag{6.17}$$

Since the above equation must hold as $b \to 0$, this implies, that

$$\frac{d^2 x_0}{dt^2} + \omega^2 x_0 = 0;$$

$$x_0(0) = A, \qquad \frac{dx_0}{dt}(0) = 0;$$

$$\frac{d^2 x_1}{dt^2} + \omega^2 x_1 = -x_0^3;$$

$$x_1(0) = 0, \qquad \frac{dx_1}{dt}(0) = 0.$$

The solution of the above equations is

$$x_0(t) = A \cos(\omega t), \tag{6.18}$$

$$x_1(t) = \frac{A^3}{32\omega^2} [\cos 3\omega t - \cos \omega t] - \left\{ \frac{3tA^3}{8\omega} \right\} \sin \omega t. \tag{6.19}$$

The $\{\cdot\}$ term is called a secular term, and its unboundedness is troublesome. If the problem were linear, it might have a periodic motion of the form

$$x(t) = A \cos \omega_0 t. \tag{6.20}$$

If we use perturbation notation, this may be represented as

$$x(t) = A \cos(\omega_0 t + \omega_1 t). \tag{6.21}$$

A power-series expansion of this yields

$$\cos(\omega_0 t + \omega_1 t) = \cos \omega_0 t \cos \omega_1 t - \sin \omega_0 t \sin \omega_1 t$$

$$= \left\{ 1 - \frac{(\omega_1 t)^2}{2!} + \frac{(\omega_1 t)^4}{4!} + \cdots \right\} \cos \omega_0 t$$

$$- \left\{ \omega_1 t - \frac{(\omega_1 t)^3}{3!} + \cdots \right\} \sin \omega_0 t.$$

At this point, the above equation also *appears* to contain secular terms! In general, unlike linear systems, the frequency of a nonlinear system is not fixed by the differential equation alone. It may have several component values, including the one assigned as ω_0 in the previous solution. If the system frequency is assumed to have multiple frequencies, we may write it in the same manner as was done for the solution variable, x. Thus, the frequency will take on the form of a power series in b,

$$\omega^2 = \omega_0^2 + b\omega_1^2 + b^2 \omega_2^2 + \cdots, \tag{6.22}$$

6.1. Method of Perturbation

where ω_0 is the natural frequency found earlier, $\omega_0^2 = a$. This assumed form for the frequency modifies the solution completely, and one obtains the following set of equations in a manner similar to that used before:

$$\frac{d^2 x_0}{dt^2} + \omega^2 x_0 = 0,$$

$$\frac{d^2 x_1}{dt^2} + \omega^2 x_1 = \omega_1^2 x_0 - x_0^3,$$

$$\frac{d^2 x_2}{dt^2} + \omega^2 x_2 = \omega_1^2 x_1 - \omega_2^2 x_0 - 3 x_0 x_1,$$

$$\vdots$$

which have the initial conditions $x_0(0) = A$, $dx_0(0)/dt = 0$, $x_1(0) = x_2(0) = \cdots = 0$, $dx_1(0)/dt = dx_2(0)/dt = \cdots = 0$. The solution of this set of equations proceeds from top to bottom and yields

$$x_0 = A \cos(\omega t), \tag{6.23}$$

$$x_1 = P_1 \cos(\omega t) + Q_1 \sin(\omega t) + t \left(\frac{\omega_1^2 A}{2\omega} - \frac{3 A^3}{8\omega} \right) \sin(\omega t) + \frac{A^3}{32 \omega^2} \cos(3\omega t). \tag{6.24}$$

It is obvious that this solution still contains a secular term, even though the actual system is bounded with respect to time. In studying Eq. (6.24), it is possible for us to choose the coefficient of the secular term so that it vanishes, that is,

$$\omega_1^2 = \frac{3 A^2}{4}. \tag{6.25}$$

With the frequency terms complete, $\omega^2 = \omega_0^2 + 3 b A^2 / 4$, we now may write the first two terms of the solution,

$$x_0 = A \cos(\omega t), \tag{6.26}$$

$$x_1 = -\frac{A^3}{32 \omega^2} [\cos(\omega t) - \cos(3 \omega t)], \tag{6.27}$$

yielding the complete solution function as the sum of these two terms:

$$x = \left(A - \frac{b A^3}{32 \omega^2} \right) \cos(\omega t) + \frac{b A^3}{32 \omega^2} \cos(3 \omega t). \tag{6.28}$$

Note the amplitude dependence on the frequency, ω, of the solution. This is one of the unusual properties of nonlinear systems which has no equal in linear systems. In addition, the amplitude of the solution is not linearly related to the boundary condition, A, as it would be if this were a linear system. The inclusion of additional terms would, of course, yield additional frequency correction terms ($\omega_3, \omega_4, \ldots$).

To extend the results of this analysis into a numerical example, consider the following values for the differential equation defining the motion of a bead on a wire:

$$\frac{d^2 x}{dt^2} + 10 x + 100 x^3 = 0, \quad x(0) = X_0. \tag{6.29}$$

TABLE 6.1. Comparison of Perturbation and Exact Solutions

X_0	T_{NL}/T_L		Error %
	Perturbation	Exact	
0.1	0.965	0.965	0.0
0.2	0.878	0.878	0.0
0.4	0.678	0.676	0.3
0.6	0.525	0.521	0.8
0.8	0.420	0.415	1.2
1.0	0.347	0.347	0.0

For the perturbation method to work, the following relation must be valid:

$$|10X_0| \gg |100X_0^3| \tag{6.30}$$

or

$$X_0^2 \ll 0.10. \tag{6.31}$$

The results of this analysis for varying initial conditions are compared to the exact solution and are tabulated in Table 6.1, where T_{NL} = period for the nonlinear solution and T_L = period for the linear solution.

Thus, the original assumption that $X_0^2 \ll 0.1$ was very conservative. In this instance, it is possible to extend the analysis beyond the predicted limiting value. The errors shown in Table 6.1 grow throughout the period, but then drop to zero because of the inherent accuracy of the frequency of the harmonic solution.

EXAMPLE 6.3: DUFFING EQUATION

In Example 6.2, the governing equation was homogeneous and the initial condition was nonhomogeneous. A variation of this equation in nonhomogeneous form and having a harmonic forcing function is termed the Duffing equation. This is a classic equation in nonlinear control systems:

$$\frac{d^2x}{dt^2} + ax + bx^3 = bF_0\cos(\omega t), \qquad x(0) = A, \qquad \left.\frac{dx}{dt}\right|_{t=0} = 0. \tag{6.32}$$

We set $x(t) = x_0(t) + bx_1(t) + \cdots$ and $\omega^2 = a + b\omega_1^2$, and proceed as before. The differential equation then can be reduced to the set of equations

$$\frac{d^2x_0}{dt^2} + \omega^2 x_0 = 0, \tag{6.33}$$

$$\frac{d^2x_1}{dt^2} + \omega^2 x_1 = \omega_1^2 x_0 - x_0^3 + bF_0\cos(\omega t), \tag{6.34}$$

with nonzero initial conditions only on the zeroth solution, $x_0(0) = A, dx_0/dt(0) = 0$. To avoid the secular term, we must pick the following value for the frequency:

$$\omega_1^2 = \frac{3A^2}{4} - \frac{F_0}{A}, \tag{6.35}$$

from which we obtain the solution

$$\omega^2 = a + \frac{3bA^2}{4} - \frac{bF_0}{A}, \tag{6.36}$$

$$x(t) = \left(A - \frac{bA^3}{32\omega^2}\right)\cos(\omega t) + \frac{bA^3}{32\omega^2}\cos(3\omega t). \tag{6.37}$$

The addition of more terms yields additional frequency correction terms and correction terms. Notice that for every amplitude, A, there are two possible frequency terms resulting from Eq. (6.36).

6.2 ITERATION TECHNIQUE

This solution technique is similar in approach to that used in the perturbation method. In both cases, a zeroth-order solution is used to represent the gross behavior of the system and correction terms are sought as solutions to secondary differential equations. In this method, however, the zeroth-order solution is obtained from the linear, homogeneous form of the differential equation. This solution then is inserted into the differential equation, thereby effecting a change in the nonlinear terms. The resulting differential equation is solved to obtain a solution for the first-order correction. This is again returned to the differential equation and the method is applied iteratively. We again seek to illuminate the method by applying it to a simple problem.

Consider the differential equation with a nonlinear term, $f(x)$:

$$\frac{d^2x}{dt^2} + ax + f(x) = F\cos(\omega t). \tag{6.38}$$

As a first approximation, it is assumed that some type of periodic output would result since the forcing function is periodic. The differential equation is rewritten as

$$\frac{d^2x}{dt^2} + \omega^2 x = F\cos(\omega t) + (\omega^2 - a)x - f(x). \tag{6.39}$$

The zeroth-order solution is generated by solving the differential equation

$$\frac{d^2x_0}{dt^2} + \omega^2 x_0 = 0, \tag{6.40}$$

and all of the initial conditions are imposed on this zeroth-order component of the solution. A first-order correction to this solution is obtained by iterative solution of Eq. (6.39) using the zeroth-order solution in place of the original (assumed) harmonic solution function. The differential equation thus becomes

$$\frac{d^2x_1}{dt^2} + \omega^2 x_1 = F\cos(\omega t) + (\omega^2 - a)x_0 - f(x_0). \tag{6.41}$$

The solution to the above equation then is used to generate the differential equation for a second-order solution:

$$\frac{d^2 x_2}{dt^2} + \omega^2 x_2 = F\cos(\omega t) + (\omega^2 - a)x_1 - f(x_1). \tag{6.42}$$

This is repeated until one's patience is exhausted or an answer is obtained. Unlike the perturbation method, each subsequent solution stands alone and must therefore also satisfy the initial conditions of the problem. This method converts the original nonlinear problem to an iterative solution of linear problems. Both Eqs. (6.41) and (6.42) are linear because the previous solutions, x_0 and x_1, respectively, are known functions of time. This may be expressed as

$$\frac{d^2 x_n}{dt^2} + \omega^2 x_n = F\cos(\omega t) + (\omega^2 - a)x_{n-1}(t) - f(x_{n-1}(t)) = G(t), \tag{6.43}$$

which is readily seen to be linear. Because of the rapidly increasing complexity of the algebraic operations, the use of a symbolic algebraic manipulation program is of enormous benefit in the application of this technique.

EXAMPLE 6.4: DUFFING EQUATION

The classic nonlinear Duffing equation is

$$\frac{d^2 x}{dt^2} + ax + bx^3 = F\cos(\omega t), \qquad x(0) = A, \qquad \dot{x}(0) = 0. \tag{6.44}$$

This differential equation may be rearranged to become

$$\frac{d^2 x}{dt^2} + \omega^2 x = (\omega^2 - a)x - bx^3 + F\cos(\omega t), \qquad x(0) = A, \qquad \dot{x}(0) = 0. \tag{6.45}$$

To begin the iteration solution method, the first approximation to the solution is found from the homogeneous form of the Duffing equation,

$$\frac{d^2 x_0}{dt^2} + \omega^2 x_0 = 0, \qquad x_0(0) = A, \qquad \dot{x}_0(0) = 0. \tag{6.46}$$

The solution to this linear, homogeneous equation is a simple harmonic,

$$x_0 = A\cos(\omega t). \tag{6.47}$$

The first iterative improvement on this elementary solution then is found from the differential equation

$$\frac{d^2 x_1}{dt^2} + \omega^2 x_1 = \left[(\omega^2 - a)A - \frac{3bA^3}{4} - F\right]\cos(\omega t) - \frac{bA^3}{4}\cos(3\omega t), \tag{6.48}$$
$$fx_1(0) = A, \qquad \dot{x}_1(0) = 0,$$

or, setting the bracketed terms to zero to avoid a secular term, we obtain an equation for the frequency, ω,

$$\omega^2 = a + \frac{3bA^2}{4} - \frac{F}{A}, \tag{6.49}$$

yielding the first iterative solution

$$x_1 = \left(A - \frac{bA^3}{32\omega^2}\right)\cos(\omega t) + \frac{bA^3}{32\omega^2}\cos(3\omega t). \tag{6.50}$$

Only the forcing function term, $\cos(\omega t)$, yields a particular solution to Eq. (6.48); the term containing $\cos(3\omega t)$ does not. The results for both the frequency of oscillation and the response are the same as obtained using the perturbation method [Eqs. (6.37) and (6.36)]. This process can be repeated as needed to further refine and improve the solution.

6.3 POWER-SERIES METHOD

In this method, the form of the differential equation is

$$\frac{d^2x}{dt^2} + ax + bx^3 = F\cos(\omega t), \tag{6.51}$$

subject to the boundary conditions $x(0) = A$, $dx/dt(0) = 0$. A solution is assumed in the form

$$x(t) = A_1\cos(\omega t) + A_3\cos(3\omega t). \tag{6.52}$$

The solution form assumed here is based on the knowledge derived from previous solutions. By defining $\beta = \omega t$, the differential equation becomes

$$\frac{d^2x}{dt^2} = -\omega^2[A_1\cos\beta + 9A_3\cos3\beta],$$
$$ax = a[A_1\cos\beta + A_3\cos3\beta],$$
$$bx^3 = \frac{b}{4}\{A_1^3[3\cos\beta + \cos3\beta] + 3A_1^2A_3[\cos\beta + 2\cos3\beta + \cos5\beta]$$
$$+ 3A_1A_3^2[2\cos\beta + \cos5\beta + \cos7\beta] + A_3^3[3\cos3\beta + \cos9\beta]\}.$$

Euler's expansions for trigonometric functions are employed next:

$$\cos^3 x = \tfrac{3}{4}\cos x + \tfrac{1}{4}\cos 3x. \tag{6.53}$$

This may be substituted into the governing equation to obtain two equations, one for $\cos\beta$ and one for $\cos3\beta$. If all higher-order terms are neglected, the following is obtained for the coefficient of $\cos\beta$:

$$\omega^2 = \left[a + \frac{3bA_1^2}{4} - \frac{F}{A_1}\right] + \frac{3bA_1^2}{4}\left(\frac{A_3}{A-1}\right)\left[1 + 2\frac{A_3}{A_1}\right]. \tag{6.54}$$

To reduce the complexity of the problem somewhat, we approximate the frequency by assuming that $A_3/A_1 \ll 1$ and obtain

$$\omega^2 = a + \frac{3bA_1^2}{4} - \frac{F}{A_1}. \tag{6.55}$$

The coefficient for $\cos 3\beta$ yields a separate equation,

$$(a - 9\omega^2)A_3 + \frac{bA_1^3}{4} + \frac{3bA_1^2 A_3}{2} + \frac{3bA_3^3}{4} = 0, \tag{6.56}$$

or

$$A_3 = \frac{\frac{bA_1^3}{4}}{9\omega^2 - a - 1.5bA_1^2 - 0.75bA_3^2}. \tag{6.57}$$

Again, imposing the condition that $A_3/A_1 \ll 1$, and including the expression for ω^2, we obtain

$$A_3 = \frac{bA_1^3}{\left[36\omega^2 - 4a - 6bA_1^2\right]}. \tag{6.58}$$

EXAMPLE 6.5: NONLINEAR SPRING-MASS PROBLEM

Consider the problem given by

$$\frac{d^2x}{dt^2} + 10x + 100x^3 = 0, \quad x(0) = 1, \quad \frac{dx}{dt}(0) = 0. \tag{6.59}$$

Following the recipe given above, we assume that

$$x = A_1 \cos(\omega t) + A_3 \cos(3\omega t). \tag{6.60}$$

The initial conditions and the result of Eq. (6.58) may be applied here, and they result in

$$x(0) = 1 = A_1 + A_3, \tag{6.61}$$

$$A_3 = \frac{100 A_1^3}{320 + 2100 A_1^2}. \tag{6.62}$$

Solving these two equations simultaneously yields

$$A_1 = 0.9607,$$

$$A_3 = 0.0393.$$

Equation (6.55) produces the characteristic frequency of the system, $\omega = 8.903$. The period of motion then is computed as

$$T = 2\pi/\omega = 0.706. \tag{6.63}$$

The period obtained by the exact solution (elliptic integral) is

$$T_{\text{exact}} = 0.695. \tag{6.64}$$

The magnitude of the nonlinear terms and their period therefore has been resolved to a reasonable approximation. This is valuable information for the solution and understanding of a difficult problem.

6.3.1 Duffing Equation

Equation (6.65) is a classic equation in nonlinear systems known as the Duffing equation:

$$\frac{d^2x}{dt^2} + c\frac{dx}{dt} + ax + bx^3 = F\cos(\omega t + \phi). \tag{6.65}$$

In this equation, the phase-angle term, ϕ, is inserted to make the solution simpler in form, and F is the ratio of the amplitude of the force to the mass of the system. In addition to describing the motion of a mass on a nonlinear spring, it also describes the response of a nonlinear electrical capacitive-inductor series circuit driven by a sinusoidal voltage. Under the proper conditions, Duffing systems exhibit a response in which the output frequencies are subharmonics of the driving frequency. The subharmonic that is easiest to attain is of the form

$$x(t) = A_1 \cos(\omega t) + A_3 \cos(3\omega t). \tag{6.66}$$

The forcing function may be expanded as

$$F\cos(\omega t + \phi) = F(\cos\omega t \cos\phi - \sin\omega t \sin\phi), \tag{6.67}$$

and the differential equation may be grouped to equate the cosine and sine terms on each side of the equation. Assuming that $A_3 \ll A_1$, this equation yields the cosine and sine terms

$$\begin{aligned}\cos(\omega t): & \quad \left(a - \omega^2 + \frac{3bA_1^2}{4}\right)A_1 = F\cos\phi, \\ \sin(\omega t): & \quad \omega c A_1 = F\sin\phi.\end{aligned} \tag{6.68}$$

Squaring these two equations and adding yields

$$A_1 = F\left[\left(a - \omega^2 + \frac{3bA_1^2}{4}\right)^2 + \omega^2 c^2\right]^{-1/2}, \tag{6.69}$$

which yields

$$A_1^2 = \frac{2}{3b}\left[\omega^2(a - c^2) + \sqrt{a^2 - 2a\omega^2(1-c^2) + \omega^4(1-c^2)^2 + 3F^2 b}\right]. \tag{6.70}$$

It is salient that the amplitude [Eq. (6.70)] is a function of the excitation frequency, ω, as well as the parameters in the differential equation. Careful examination of Eq. (6.70) also reveals that the amplitude is not linearly related to the amplitude of the forcing function. The phase angle, ϕ [Eq. (6.68)], has similar dependencies. Unlike linear systems, the phase angle is directly dependent on the amplitude of the forcing function, F. These characteristics are typical of nonlinear systems, as noted in Chapter 2, Section 2.1.

An interesting phenomenon of the Duffing equation is called *jump resonance*. This is depicted in Fig. 6.2. As noted in the figure, there is a region of the possible amplitudes that is double valued. For the case of $b > 0$, if one starts at a low frequency and begins to increase the frequency, the amplitude builds to a maximum at A and then falls to B. If the frequency continues to increase, the amplitude suddenly jumps down to C and continues to fall. This same phenomenon can be experienced when driving over a dirt road that

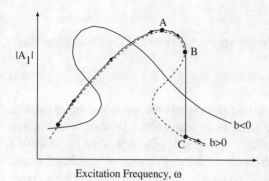

Figure 6.2. Jump resonance phenomenon of a nonlinear system.

has a washboard-like surface. As the vehicle speed increases, the bumpiness of the ride increases, followed by a sudden and dramatic drop in the amplitude of the oscillations of the suspension. The reverse phenomenon occurs during speed reduction, but the transition is at a lower speed.

EXAMPLE 6.6: VEHICLE SUSPENSION SYSTEM

Consider the simple model of an automobile suspension system as shown in Fig. 6.3. The order of magnitude of the natural frequencies of oscillation of the suspension is approximately 1 Hertz if the axle is fixed in space and approximately 10 Hertz if the car body is fixed. Consider the case in which the inertia of the car body is such that y is approximately constant (fixed body). For this case, the force-compression character of the tire is depicted in Fig. 6.4. The equation of motion for the system shown is

$$m\ddot{x} + f(x - y) = 0. \tag{6.71}$$

If we define $z = (x - y)$, then this equation becomes

$$m\ddot{z} + f(z) = -m\ddot{y}. \tag{6.72}$$

If we assume that y is chosen to be the input forcing function and is of the form $y = A\cos\omega t$, then the differential equation becomes

$$m\ddot{z} + f(z) = mA\omega^2 \cos\omega t. \tag{6.73}$$

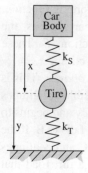

Figure 6.3. Simplified model of an automobile suspension system.

6.3. Power-Series Method

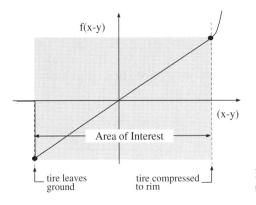

Figure 6.4. Force-compression of an automobile tire.

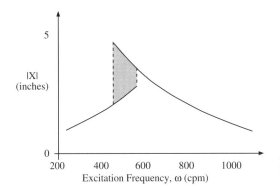

Figure 6.5. Jump resonance of an automobile suspension system.

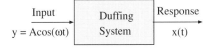

Figure 6.6. Generalized Duffing system.

The function shown in Fig. 6.4 is approximated by the polynomial

$$f(z) = az + bz^3. \tag{6.74}$$

Applying the power-series method to this problem yields the results shown in Fig. 6.5 (adapted from Schilling [40]). The jump resonance phenomenon is clearly evident.

6.3.2 Generalized Duffing Equation

In general, for an input y as depicted in Fig. 6.6, the system has a response x. This response may have higher harmonics ($A_n \cos(n\omega t)$) as well as subharmonics ($A_{1/n} \cos(\omega t/n)$).

Consider the undamped form of the Duffing equation:

$$\frac{d^2x}{dt^2} + ax + bx^3 = F \cos(\omega t + \phi). \tag{6.75}$$

The existence of forcing functions that are multiples of the fundamental frequency of oscillation of a system provide the potential for the production of subharmonics. To discern the conditions for this to occur, a solution is assumed of the form

$$x = A_{1/3} \cos\left(\frac{\omega t}{3}\right) + A_1 \cos(\omega t). \tag{6.76}$$

This can be substituted into the differential equation and grouped into terms having like cosine terms. This yields

$$\cos\left(\frac{\omega t}{3}\right): \left[\left(a - \frac{\omega^2}{9}\right) + \frac{3b}{4}(A_{1/3}^2 + A_1 A_{1/3} + 2A_1^2)\right] A_{1/3} = 0, \tag{6.77}$$

$$\cos(\omega t): \left[\left(a - \omega^2\right)A_1 + \frac{b}{4}(A_{1/3}^3 + 6A_1 A_{1/3}^2 + 3A_1^3)\right] - F = 0. \tag{6.78}$$

By assuming that $A_{1/3} \neq 0$, it is possible to solve the remaining two equations for

$$A_{1/3} = -\frac{1}{2}\left[A_1 \pm \sqrt{\frac{16}{27b}(\omega^2 - 9a) - 7A_1^2}\right]. \tag{6.79}$$

Since $A_{1/3}$ must be real, the radical must be ≥ 0. This infers a lower bound for ω, below which $A_{1/3}$ may not exist. Thus,

$$\omega^2 \geq 9\left(a + \frac{21 A_1^2}{16}b\right). \tag{6.80}$$

Therefore, when $\omega^2 < 9a$, linear resonance results with no subharmonics. Above $\omega^2 \geq 9a$, the oscillations include at least one subharmonic. When the harmonic first appears, it suddenly takes on the amplitude $A_{1/3} = -A_1/2$. At this same point, the two equations may be manipulated to obtain

$$\frac{343b}{32}A_1^3 + 8aA_1 + F = 0. \tag{6.81}$$

For small values of the parameter b, the amplitude of the fundamental harmonic is

$$A_1 \approx -F/8a, \tag{6.82}$$

and the frequency of this harmonic is

$$\omega^2 \geq 9\left(a + \frac{21bF^2}{1024a^2}\right). \tag{6.83}$$

EXAMPLE 6.7: DUFFING SYSTEM WITH SUBHARMONICS

Consider the special case of $F = 1$, $a = 1$, and $b = 0.1$. The analysis presented in this section permits a computation of the fundamental and subharmonic amplitudes as a function of the frequency, ω. This is depicted in Fig. 6.7. As a practical note, subharmonics only appear in systems with very light damping. If the governing equation is reformulated in the following manner after the usual linear systems form, that is,

$$\ddot{x} + 2\xi\omega_n \dot{x} + \omega_n^2 x + bf(x) = F\cos(\omega t), \tag{6.84}$$

then subharmonics will exist only if $\xi \leq 0.01$.

6.4. Method of Harmonic Balance

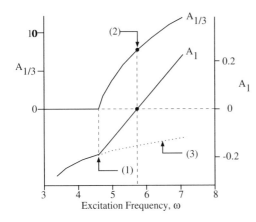

Figure 6.7. Frequency content of a Duffing system: (1) subharmonics exist below this frequency; (2) this frequency, only the subharmonic exists; (3) A_1 for a linear system.

6.4 METHOD OF HARMONIC BALANCE

This method is referred to as the method of slowly varying parameters or the method of Krylov[1] and Bogoliubov[2] [24]. It is useful for problems whose solution functions are harmonic and have amplitudes and phase angles that vary slowly with time. The solution function thus takes on the following form:

$$x(t) = a(t) \sin(\omega_0 t + \phi(t)), \tag{6.85}$$

where $a(t)$ and $\phi(t)$ vary slowly with respect to $\omega_0 t$. Thus, da/dt and $d\phi/dt$ are "small" in some sense. The solution function in Eq. (6.85) thus is assumed to be a valid solution of the general differential equation of the form

$$\ddot{x} + F(x, \dot{x}) = 0. \tag{6.86}$$

The function F then is formulated as containing linear and nonlinear terms:

$$F(x, \dot{x}) = \omega_0^2 x + \mu f(x, \dot{x}). \tag{6.87}$$

The differential equation thus takes on the form common to the method of harmonic balance:

$$\ddot{x} + \omega_0^2 x + \mu f(x, \dot{x}) = 0. \tag{6.88}$$

If the solution function of Eq. (6.85) is substituted into Eq. (6.88), the resultant becomes

$$\ddot{a} \sin(u) + 2\dot{a}(\omega_0 + \dot{\phi}) \cos(u) - a(\omega_0 + \dot{\phi})^2 \sin(u) + a\ddot{\phi} \cos(u) + a\omega_0^2 \sin(u)$$
$$\approx -\mu f[a \sin(u), a\omega_0 \cos(u)], \tag{6.89}$$

where

$$u = \omega_0 t + \phi(t), \tag{6.90}$$

[1] A.N. Krylov (1863–1945).
[2] N.N. Bogoliubov (1909–).

and where the term $a\omega_0 \cos(u)$ is an approximation of $\dot{x}$. This equation is further simplified by assuming that

$$\dot{a} \ll \omega_0 a, \quad \dot{\phi} \ll \omega_0,$$

and that all the second-order derivatives and cross products can be neglected, resulting in

$$2\dot{a}\omega_0 \cos(u) - 2a\omega_0\dot{\phi} \sin(u) = -\mu f[a \sin(u), a\omega_0 \cos(u)]. \tag{6.91}$$

Since the function $f[a \sin(u), a\omega_0 \cos(u)]$ is harmonic, it is assumed that it can be expanded as a Fourier series,

$$f[a \sin(u), a\omega_0 \cos(u)] = f_s \sin(u) + f_c \cos(u) + R, \tag{6.92}$$

where f_s is the coefficient of the $\sin(u)$ term, f_c is the coefficient of the $\cos(u)$ term, and R represents all remaining terms. Equating coefficients of like terms yields

$$\dot{a} = -\frac{\mu f_c}{2\omega_0}, \tag{6.93}$$

$$\dot{\phi} = \frac{\mu f_s}{2a\omega_0}. \tag{6.94}$$

The coefficients f_s, f_c are found by using the orthogonality of the Fourier harmonic functions,

$$\int_0^{2\pi} f \sin(u) \, du = \int_0^{2\pi} f_s \sin^2(u) \, du + \int_0^{2\pi} f_c \cos(u) \sin(u) \, du$$

$$+ \int_0^{2\pi} f_R \sin(u) \, du = \pi f_s, \tag{6.95}$$

$$\int_0^{2\pi} f \cos(u) \, du = \int_0^{2\pi} f_c \cos^2(u) \, du + \int_0^{2\pi} f_s \cos(u) \sin(u) \, du$$

$$+ \int_0^{2\pi} f_R \cos(u) \, du = \pi f_c, \tag{6.96}$$

to obtain the relationships

$$f_s = \frac{1}{\pi} \int_0^{2\pi} f[a \sin(u), a\omega_0 \cos(u)] \sin(u) \, du, \tag{6.97}$$

$$f_c = \frac{1}{\pi} \int_0^{2\pi} f[a \sin(u), a\omega_0 \cos(u)] \cos(u) \, du. \tag{6.98}$$

Finally, the time dependence of the parameters $a(t), \phi(t)$ are solutions to the equations

$$\dot{a} = -\frac{\mu}{2\pi \omega_0} \int_0^{2\pi} f \cos(u) \, du \tag{6.99}$$

and

$$\dot{\phi} = \frac{\mu}{2a\pi \omega_0} \int_0^{2\pi} f \sin(u) \, du. \tag{6.100}$$

6.4. Method of Harmonic Balance

Equations (6.99) and (6.100) are two coupled linear differential equations for ascertaining the time dependence of the amplitude and phase, respectively, of the harmonic response.

EXAMPLE 6.8: VAN DER POL OSCILLATOR

In 1934, van der Pol [51][3] studied the electrical oscillations of a vacuum-tube circuit. The governing equation, termed the van der Pol oscillator equation, for this system is given as

$$\ddot{x} - \mu(1 - x^2)\dot{x} + x = 0. \tag{6.101}$$

This can be compared to the form of Eq. (6.88) to find that

$$\omega^2 = 1, \tag{6.102}$$

$$f(x, \dot{x}) = -(1 - x^2)\dot{x}. \tag{6.103}$$

Thus we can solve for the parameters a, ϕ from the integro-differential equations

$$\dot{a} = +\frac{\mu}{2\pi} \int_0^{2\pi} [(1 - a^2 \sin^2(u))a\omega_0 \cos(u)] \cos(u)\, du$$
$$= \frac{\mu}{2} \left(a - \frac{a^3}{4} \right), \tag{6.104}$$

$$\dot{\phi} = -\frac{\mu}{2\pi} \int_0^{2\pi} [(1 - a^2 \sin^2(u))a\omega_0 \cos(u)] \sin(u)\, du$$
$$= 0. \tag{6.105}$$

Integration of these two yields

$$\phi = \phi_0, \tag{6.106}$$

$$a(t) = \frac{a_0 e^{\mu t/2}}{\sqrt{1 + a_0^2 \left(e^{\mu t} - 1 \right)/4}}. \tag{6.107}$$

The phase shift is therefore a constant, fixed by the initial conditions. In addition, the amplitude has an initial transient that decays to a limiting value given as

$$\lim_{t \to \infty} a(t) = 2. \tag{6.108}$$

EXAMPLE 6.9: MASS-DAMPER WITH A CUBIC SPRING

Consider a nonlinear intertial-elastic system whose governing differential equation is

$$\ddot{x} + 2C\dot{x} + kx + bx^3 = 0. \tag{6.109}$$

---

[3] B. van der Pol (1889–1959).

By inspection, the governing equation can be cast into the form of the method of harmonic balance, and the following relevant terms are found:

$$\omega_0^2 = k, \tag{6.110}$$

$$\mu f(x, \dot{x}) = 2c\dot{x} + bx^3. \tag{6.111}$$

Using Eqs. (6.99) and (6.100), it is possible to obtain differential equations for the amplitude and phase as

$$\begin{aligned}\dot{a} &= -\frac{1}{2\pi\omega_0}\int_0^{2\pi}[2Ca\omega_0\cos(u) + ba^3\sin^3(u)]\cos(u)\,du \\ &= -Ca,\end{aligned} \tag{6.112}$$

$$\begin{aligned}\dot{\phi} &= \frac{1}{2\pi a\omega_0}\int_0^{2\pi}[2Ca\omega_0\cos(u) + ba^3\sin^3(u)]\sin(u)\,du \\ &= \frac{3ba^2}{8\omega_0}.\end{aligned} \tag{6.113}$$

These are readily integrated to obtain

$$a = a_0 e^{-Ct},$$

$$\phi = \phi_0 - \frac{3ba_0^2}{16C\omega_0}(e^{-2Ct} - 1). \tag{6.114}$$

Note also that

$$\lim_{t\to\infty} a = 0. \tag{6.115}$$

This solution has an amplitude that is only a linear function of the initial conditions and the parameters of the system. The phase, however, also depends on the initial conditions, but in a nonlinear manner.

6.4.1 Equivalent Linear Equation

This section includes a derivation of the relationships for damping ratio and frequency of a nonlinear system when cast into the form of an equivalent linear system. If, in the preceding section, the nonlinear term $\mu f(x, \dot{x})$ is replaced by

$$\mu f(x, \dot{x}) = F(x, \dot{x}) - \omega^2 x, \tag{6.116}$$

this produces a somewhat different formulation for $\dot{a}$, i.e.,

$$\dot{a} = -\frac{1}{2\pi\omega_0}\int_0^{2\pi}[F\{a\sin(u), a\omega\cos(u)\} - \omega^2 a\sin(u)]\cos(u)\,du, \tag{6.117}$$

and the frequency becomes

$$\begin{aligned}\omega^2 &= (\omega_0 + \dot{\phi})^2, \\ &= \omega_0^2 + 2\omega_0\dot{\phi} + \dot{\phi}^2, \\ &\approx \omega_0^2 + 2\omega_0\dot{\phi},\end{aligned} \tag{6.118}$$

6.4. Method of Harmonic Balance

or, introducing the integral form for $\dot{\phi}$,

$$\omega^2 = \omega_0^2 + \frac{1}{\pi a} \int_0^{2\pi} \left[F - \omega_0^2 a \sin u \right] \sin u \, du,$$

$$= \frac{1}{\pi a} \int_0^{2\pi} F[a \sin(u), a\omega_0 \cos(u)] \sin(u) \, du. \tag{6.119}$$

The two differential equations can be compared:

$$\ddot{x} = -2\xi\omega\dot{x} - \omega^2 x,$$
$$\ddot{x} = F[x, \dot{x}]. \tag{6.120}$$

Inserting the basic solution behavior from Eq. (6.85) into the above equation and neglecting higher-order terms yields

$$2\dot{a}(\omega_0 + \dot{\phi})\cos(u) - a(\omega_0 + \dot{\phi})^2 \sin(u) = 2\xi\omega[\dot{a}\sin(u) + a(\omega_0 + \dot{\phi})\cos(u)]$$
$$+ a\omega^2 \sin(u) = 0.$$

Gathering terms and equating each of the coefficients of the sine and cosine terms to zero yields

$$\sin(u): \quad -a(\omega_0 + \dot{\phi})^2 + 2\xi\omega\dot{a} + a\omega^2 = 0,$$
$$\cos(u): \quad 2\dot{a}(\omega_0 + \dot{\phi}) + 2\xi\omega a(\omega_0 + \dot{\phi}) = 0.$$

Assuming $(\dot{a})^2$ is small, this yields,

$$2\xi\omega = -\frac{2\dot{a}(\omega_0 + \dot{\phi})}{a(\omega_0 + \dot{\phi})}$$
$$\approx -\frac{2\dot{a}}{a}.$$

Since $\dot{a}^2/a$ is assumed to be small,

$$\omega^2 = (\omega_0 + \dot{\phi})^2 + 2\dot{a}^2/a \approx (\omega_0 + \dot{\phi})^2. \tag{6.121}$$

If the above values are substituted into the differential equation, a form of this equation with slowly varying damping ratio and frequency can be found to be

$$\ddot{x} + \left(-\frac{2\dot{a}}{a}\right)\dot{x} + (\omega_0 + \dot{\phi})^2 x = 0. \tag{6.122}$$

EXAMPLE 6.10: VAN DER POL OSCILLATOR

The governing equation for this classic oscillator system is given as

$$\ddot{x} - \mu(1 - x^2)\dot{x} + x = 0. \tag{6.123}$$

Assuming the response to have a slowly varying amplitude and phase, we can invoke Eqs. (6.99) and (6.100) to find that the phase angle is constant, $\dot{\phi} = 0$, and thus $\omega^2 = \omega_0^2 = 1$. The differential equation for the amplitude is

$$\dot{a} = \frac{\mu a}{2}\left(1 - \frac{a^2}{4}\right). \tag{6.124}$$

This produces the equivalent linear equation

$$\ddot{x} - \mu\left(1 - \frac{a^2}{4}\right)\dot{x} + x = 0. \tag{6.125}$$

Note that, for this system, where the forcing function is zero and the damping is nonzero, there exist limit cycles. At the limit, the coefficient of the damping term vanishes and the system continues to cycle at $\omega = \omega_0 = 1$. It is left to the reader to verify that the amplitude of the limit cycle is $a(\infty) = 2$.

6.5 GALERKIN'S METHOD

This method was developed by Galerkin,[4] a turn-of-the-century Russian mathematician, to aid in solving elasticity problems. Its accuracy and breadth of application to nonlinear problems have resulted in widespread use of this technique. It is termed a *weighted residue method* and utilizes the orthogonality property of functions to achieve a solution. The accuracy of this technique depends solely on the ability of the user to guess the appropriate functions for use in representing the solution function.

Consider the following representation of a differential equation to which a solution is sought:

$$L(x) = 0, \tag{6.126}$$

where L is a general operator representing derivatives and functions of derivatives. A solution is hypothesized (approximating function) of the form

$$\tilde{x} = \sum_{i=1}^{n} C_i \phi_i(t), \tag{6.127}$$

where C_i are constants and ϕ_i are functions chosen to satisfy the boundary conditions. Note that *each* ϕ_i must satisfy these conditions. Since these are assumed solutions, there will be some residue or error, ϵ, and the differential equation becomes

$$L(\tilde{x}) = \epsilon \neq 0. \tag{6.128}$$

The n equations necessary to evaluate the unknown coefficients are found by setting

$$\int_D \phi_i(t)\epsilon \, dt = 0, \quad i = 1, 2, \ldots, n, \tag{6.129}$$

---

[4] Boris Grigorievich Galerkin (1871–1945).

6.5. Galerkin's Method

where D is the domain where the solution is desired. The domain where the solution should be most accurate is chosen arbitrarily, thereby adding to the information one may supply from expert knowledge of the problem. Although it is not essential, the domain normally includes the boundaries where the initial conditions have been specified.

These relationships represent the mathematical requirement that the complete set of solution functions will exhibit the property of orthogonality over the domain of the problem. This is defined analytically as

$$\int_D \phi_i \phi_j \, dt = \begin{cases} 0 & i \neq j \\ C & i = j \end{cases}. \tag{6.130}$$

Therefore, there will be an integral relationship for each of the solution functions chosen. This produces an equal number of equations and unknown coefficients.

EXAMPLE 6.11: VERTICAL MOTION OF A HELICOPTER

A simplified model of the vertical motion of a helicopter without body drag and starting from rest is given as

$$\frac{dV}{dT} + V = 1, \qquad V(0) = 0. \tag{6.131}$$

In this problem, $L = d/dt[\,] + [\,] - 1$, and the approximating function must be chosen that will have the proper behavior. By inserting the initial condition into the differential equation, it is seen that the acceleration, dV/dt, is 1. When the helicopter reaches equilibrium and the acceleration becomes zero, it is seen that the normalized velocity reaches a maximum of 1. This gives some indication of what the solution function should be like; subsequently, we choose an approximating function that will have this same character, such as

$$\tilde{V} = C_0 + C_1 e^{-T}. \tag{6.132}$$

When the boundary condition is applied, the approximating function becomes

$$\tilde{V} = C_0[1 - e^{-T}] \tag{6.133}$$

and the differential equation reduces to

$$C_0 e^{-T} + C_0[1 - e^{-T}] - 1 = \epsilon = C_0 - 1. \tag{6.134}$$

Integration of Eq. (6.134) over the domain $t[0, \infty]$ with respect to the approximating function e^{-T} yields

$$\int_0^\infty \epsilon e^{-T} dT = \int_0^\infty [C_0 - 1] e^{-T} dT, \tag{6.135}$$

resulting in the solution $C_0 = 1$, or

$$\tilde{V} = 1 - e^{-T}. \tag{6.136}$$

This is the exact solution to Eq. (6.131).

In Example 6.11, the exact solution was obtained only because we guessed the correct approximating function. One of the assets of this method is that it does provide a means of integrating knowledge of what the solution function should be like. To extend the method to more than a single approximating function, a second example is introduced that is also amenable to exact solution.

EXAMPLE 6.12: LINEAR DAMPED MUSCLE PROBLEM

Consider the force balance of a damped muscle. Inertia is neglected, and the force exerted by the muscle decreases with the difference between the joint angle and its resting angle, $x = \theta - \theta_0$. The resulting free motion is described by the following differential equation and initial condition:

$$\frac{dx}{dt} - x = 0, \qquad x(0) = 1. \tag{6.137}$$

In this problem, $L = [d/dt - 1]$, and a solution function is chosen that will be representative of the expected behavior of the system. In this example, the solution is assumed to behave functionally as a power series:

$$\tilde{x} = C_0 + C_1 t + C_2 t^2 + C_3 t^3. \tag{6.138}$$

Applying the initial condition infers that $C_0 = 1$. Inserting this into the differential equation results in

$$\epsilon = (C_1 + 2C_2 t + 3C_3 t^2) - (1 + C_1 t + C_2 t^2 + C_3 t^3), \tag{6.139}$$

or

$$\epsilon = (C_1 - 1) + (2C_2 - C_1)t + (3C_3 - C_2)t^2 - C_3 t^3. \tag{6.140}$$

The lower end of the domain, $t = 0$, is already established and the upper bound on t is arbitrarily chosen to be $t = 1$. The domain is thus $0 \leq t \leq 1$, and the Galerkin method is applied, yielding the orthogonality conditions

$$\int_0^1 \epsilon t\, dt = 0 \longrightarrow 10C_1 + 25C_2 + 33C_3 = 30, \tag{6.141}$$

$$\int_0^1 \epsilon t^2\, dt = 0 \longrightarrow 5C_1 + 18C_2 + 26C_3 = 20, \tag{6.142}$$

$$\int_0^1 \epsilon t^3\, dt = 0 \longrightarrow 21C_1 + 98C_2 + 150C_3 = 105. \tag{6.143}$$

Solving the above equations simultaneously results in the values for the constants C_1, C_2, and C_3 of

$$C_1 = 1.034, \qquad C_2 = 0.388, \qquad C_3 = 0.302, \tag{6.144}$$

which produces the following solution function:

$$\tilde{x} = 1 + 1.034t + 0.388t^2 + 0.302t^3. \tag{6.145}$$

6.5. Galerkin's Method

TABLE 6.2. Comparison of Galerkin, Picard, and Secondary Galerkin Methods

t	Exact	Picard	Galerkin	Error[a]	Galerkin 2[b]	Error[c]
0.0	1.000	1.000	1.000	0.000	1.000	0.000
0.2	1.221	1.221	1.224	0.003	1.220	0.001
0.4	1.492	1.491	1.495	0.003	1.487	0.005
0.6	1.822	1.816	1.825	0.003	1.816	0.006
0.8	2.226	2.205	2.229	0.003	2.219	0.007
1.0	2.718	2.667	2.723	0.005	2.711	0.007

[a] Galerkin vs exact.
[b] Secondary Galerkin solution.
[c] Secondary Galerkin vs exact.

The exact solution can be shown to be $x(t) = e^t$. This result is obviously only valid over a small range of angles until the elastic forces within the muscle become important. Applying Picard's method similarly produces a power-series solution

$$\tilde{x} = 1 + t + 0.5t^2 + 0.167t^3. \tag{6.146}$$

A comparison of these three solutions is given in Table 6.2. Note that, if the correct solution had been chosen, $\tilde{x} = e^{Kt}$, Galerkin's method would have obtained the exact value of $K = 1$.

6.5.1 Secondary Boundary Conditions

The inclusion of additional constraints on the solution via additional boundary conditions may add to the accuracy of the solution. For Example 6.12, the approximating solution function is chosen as

$$\tilde{x} = 1 + C_1 t + C_2 t^2 + C_3 t^3, \tag{6.147}$$

which must satisfy the differential equation at the time $t = 0$, or

$$\frac{dx}{dt} - x = 0. \tag{6.148}$$

This condition results in $C_1 = 1.0$, not 1.034, as previously ascertained. Therefore, the solution function is now

$$\tilde{x} = 1 + t + C_2 t^2 + C_3 t^3. \tag{6.149}$$

If one proceeds as before, the results listed as Galerkin-2 in Table 6.2 are obtained. The constants C_2 and C_3 are found using the same method as before from the orthogonality conditions, or

$$\int_0^1 \epsilon t^2 \, dt = 0, \tag{6.150}$$

$$\int_0^1 \epsilon t^3 \, dt = 0, \tag{6.151}$$

yielding $C_2 = 0.4342$, and $C_3 = 0.2763$. The application of the Galerkin technique is now demonstrated through six additional examples.

EXAMPLE 6.13: NONLINEAR SPRING-DAMPER PROBLEM

Consider the following differential equation and initial condition:

$$\frac{dx}{dt} + x^2 = 0, \qquad x(0) = 1. \tag{6.152}$$

A power series is assumed for the form of the solution,

$$\tilde{x} = C_0 + C_1 t + C_2 t^2. \tag{6.153}$$

If the initial condition is imposed, one obtains

$$\tilde{x} = 1 + C_1 t + C_2 t^2. \tag{6.154}$$

The resulting residue term may be computed as

$$\epsilon = \frac{d\tilde{x}}{dt} + \tilde{x}^2$$
$$= (C_1 + 1) + (2C_2 + 2C_1)t + (2C_2 + C_1^2)t^2 + (2C_1 C_2)t^3 + C_2^2 t^4.$$

The orthogonality conditions are expressed by the following integrals:

$$\int_0^1 \epsilon t \, dt = 0, \tag{6.155}$$

$$\int_0^1 \epsilon t^2 \, dt = 0. \tag{6.156}$$

These yield some very long equations for the coefficients C_1 and C_2. It is possible to obtain estimates for these coefficients using this technique, but perhaps a better solution function could be guessed than a power series. Let us consider choosing a solution similar to that obtained from the linear equation,

$$\tilde{x} = e^{-Kt}. \tag{6.157}$$

The use of this approximating function results in the Galerkin residue

$$\epsilon = -K e^{-Kt} + e^{-2Kt}, \tag{6.158}$$

and the orthogonality condition is therefore

$$\int_0^\infty \epsilon e^{-Kt} \, dt = 0. \tag{6.159}$$

This results in the value for the constant $K = \frac{2}{3}$. The upper limit of ∞ was selected for mathematical convenience because the exponential term vanishes at this limit. The

6.5. Galerkin's Method

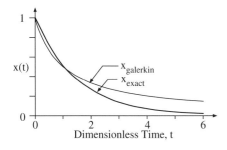

Figure 6.8. Comparison of Galerkin and exact solutions for a spring-dashpot system.

solution is therefore

$$\tilde{x} = e^{-2t/3}. \tag{6.160}$$

The relative accuracy of the Galerkin technique to the exact solution is shown in Fig. 6.8.

EXAMPLE 6.14: BEAD-ON-A-WIRE PROBLEM

Consider the motion of a bead on a wire as described in Section 3.4. Because of the geometry of the system, it has a nonlinear stress-strain relationship. The resulting nonlinear elastic terms are seen in the differential equation for the deflection, x, as a function of time, t:

$$\frac{d^2x}{dt^2} + 10x + 100x^3 = 0. \tag{6.161}$$

The initial conditions are $x(0) = A$ and $dx/dt|_{x(0)} = 0$. Because the resulting oscillation of the bead will be harmonic in nature and there are no damping terms, a Galerkin approximating function is assumed of the form

$$x(t) = c_1 \sin \omega t + c_2 \cos \omega t. \tag{6.162}$$

If the initial conditions are invoked, this reduces to

$$x(t) = A \cos \omega t. \tag{6.163}$$

The remaining difficulty is to evaluate ω. Defining $\theta = \omega t$, the orthogonality condition is then

$$\int_0^{2\pi} \epsilon \cos \theta \, d\theta = 0, \tag{6.164}$$

which is evaluated as

$$\omega_G = \sqrt{10(1 + 7.5A^2)}. \tag{6.165}$$

This results in the following estimate for the solution period:

$$P_{\text{Galerkin}} = \frac{2\pi}{\omega_G}. \tag{6.166}$$

TABLE 6.3. Comparison of Galerkin and Exact Solutions for a Bead Oscillating on a Wire

Amplitude A	Galerkin period	Exact period	Error (%)
0.0	1.987	1.987	0.000
0.2	1.743	1.745	0.115
0.4	1.340	1.348	0.593
0.6	1.033	1.045	1.148
0.8	0.825	0.837	1.434
1.0	0.682	0.693	1.587

This may be compared to the exact solution using elliptic integrals, where

$$P_{\text{Exact}} = \frac{4}{\sqrt{10 + 100A^2}} \mathcal{F}\left(k, \frac{\pi}{2}\right) \qquad (6.167)$$

and

$$k = \sqrt{\frac{100A^2}{2(10 + 100A^2)}}. \qquad (6.168)$$

These two solutions are compared in Table 6.3. Note that, as the amplitude of the initial condition, A, increases, the time course of the response changes. This is unlike linear systems in which the time constants are solely dependent on the differential equation. The method of Galerkin also predicts higher-order harmonic terms, but this is done through the proper selection of approximating solution functions. It would be a good exercise to attempt a more robust solution function for this problem. It is left to the student to try the following solution function:

$$\tilde{x} = \sum_{i=1}^{3} C_i \cos(i\omega t), \qquad (6.169)$$

and to compare this result to those given in Table 6.3.

EXAMPLE 6.15: BLASIUS EQUATION

The Blasius equation, described in Chapter 2, Section 2.5.7, is

$$f'''(\eta) + f''(\eta)f(\eta) = 0, \qquad f(0) = 0, \quad f'(0) = 0, \quad f'(\infty) = 2. \qquad (6.170)$$

To achieve a solution to this third-order ordinary nonlinear differential equation, we might attempt an approximating function of the form

$$\tilde{f}(\eta) = a_0 + a_1\eta + a_2\eta^2 + \cdots. \qquad (6.171)$$

6.5. Galerkin's Method

This power series, however, does not allow one to meet the third boundary condition, that is, $f'(\infty) = 2$. Some of the other possible approximating functions are

$$\tilde{f}'(\eta) = 2(1 - e^{-k\eta}), \tag{6.172}$$

$$\tilde{f}(\eta) = \frac{2\sin(k\eta)}{\eta}, \tag{6.173}$$

$$\tilde{f}(\eta) = \tanh(k\eta), \tag{6.174}$$

$$\tilde{f}'(\eta) = \frac{2k\eta}{1 + k\eta}. \tag{6.175}$$

Using the approximating function given by integration of Eq. (6.172), we obtain the function and its derivatives which match all of the boundary conditions:

$$\tilde{f}(\eta) = 2\eta + \frac{2}{k}(e^{-k\eta} - 1), \tag{6.176}$$

$$\tilde{f}'(\eta) = 2(1 - e^{-k\eta}), \tag{6.177}$$

$$\tilde{f}''(\eta) = 2ke^{-k\eta}, \tag{6.178}$$

$$\tilde{f}'''(\eta) = -2k^2 e^{-k\eta}. \tag{6.179}$$

These are inserted into Eq. (6.170) to yield the residue

$$\epsilon = 2e^{-k\eta}[-k^2 + 2k\eta + 2e^{-k\eta} - 2]. \tag{6.180}$$

The Galerkin orthogonality condition is thus

$$\int_0^\infty \epsilon e^{-k\eta} d\eta = 0 = -\frac{k}{2} + \frac{1}{6k}. \tag{6.181}$$

There is only a single real root and it is $k = \sqrt{3}/3$. Thus, the solution to the Blasius equation is

$$\tilde{f}(\eta) = 2\eta + \frac{6}{\sqrt{3}}\left(e^{-\sqrt{3}\eta/3} - 1\right). \tag{6.182}$$

The drag on the plate is directly proportional to the second derivative, $f''(\eta = 0)$. The commonly accepted value of this function is $f''(0) = 1.328$ and the Galerkin function of Eq. (6.182) yields $\tilde{f}''(0) = 1.154$, an error of only 13% in the *second derivative* of the approximation function. To possibly enhance our accuracy in the boundary layer, the secondary Galerkin method may be employed. It is left to the reader to evaluate this variation on the method.

EXAMPLE 6.16: COLUMN BUCKLING

A classical mechanics problem is the analysis of the buckling of an axially loaded column. This is shown pictorially in Fig. 6.9, where a load of value P is applied to a column having a flexural rigidity of EI and a length L. The resulting deflection of the

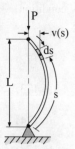

Figure 6.9. Buckling of a simple column under axial loading.

column at a distance s from the bottom of the column is described by the differential equation

$$\frac{\frac{d^2v}{ds^2}}{\sqrt{1-\left(\frac{dv}{ds}\right)^2}} + \frac{Pv}{EI} = 0. \tag{6.183}$$

Linear Problem

Under the condition that $dv/ds \ll 1$, this system behaves linearly and Eq. (6.183) reduces to

$$\frac{d^2v}{ds^2} + \frac{Pv}{EI} = 0, \qquad v(0) = v(L) = 0. \tag{6.184}$$

By defining $P/EI = K^2$, the solution to Eq. (6.184) is

$$v = A\sin(Ks) + B\cos(Ks) \tag{6.185}$$

and the boundary conditions on $v(0)$ and $v(L)$ force $B = 0$ and $K = n\pi/L$, where $n = 1, 2, 3, \ldots$. The values of n produce different modes of buckling shown in Fig. 6.10. The various modes of buckling correspond to the buckling loads given in Eq. (6.186):

$$P = \frac{n^2\pi^2 EI}{L^2}. \tag{6.186}$$

The critical buckling load is for the case $n = 1$:

$$P_{\text{crit}} = \frac{\pi^2 EI}{L^2}. \tag{6.187}$$

The solution for the deflection is thus

$$v = A\sin\left(\frac{n\pi s}{L}\right). \tag{6.188}$$

Unfortunately, the amplitude of the deflection in the center of the column, A, is unknown. Our ability to determine this quantity was effectively lost in the linearization process.

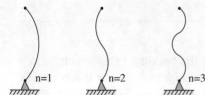

Figure 6.10. Buckling modes of an axially loaded column.

6.5. Galerkin's Method

Nonlinear Problem

If the denominator in the first term of Eq. (6.183) is expanded in a Taylor series, this function becomes

$$\frac{1}{\sqrt{1-\left(\frac{dv}{ds}\right)^2}} = \left[1 + \frac{1}{2}\left(\frac{dv}{ds}\right)^2 + \frac{3}{8}\left(\frac{dv}{ds}\right)^4 + \cdots\right]. \tag{6.189}$$

This may be inserted into Eq. (6.183) to obtain

$$\frac{d^2v}{ds^s}\left[1 + \frac{1}{2}\left(\frac{dv}{ds}\right)^2 + \frac{3}{8}\left(\frac{dv}{ds}\right)^4 + \cdots\right] + \frac{Pv}{EI} = 0. \tag{6.190}$$

To apply the Galerkin method, it is necessary to assume an approximating function and an interval over which the function is to be applied. The domain of the problem is clearly $s(0, L)$ and having already obtained a solution function from the linearized problem, it is assumed that this functional representation may be used for the nonlinear problem. This results in the approximating function

$$\tilde{v} = A \sin\left(\frac{n\pi s}{L}\right), \tag{6.191}$$

and the orthogonality condition becomes

$$\int_0^L \left(\frac{d^2\tilde{v}}{ds^s}\left[1 + \frac{1}{2}\frac{d\tilde{v}^2}{ds} + \frac{3}{8}\frac{d\tilde{v}^4}{ds} + \cdots\right] + \frac{P\tilde{v}}{EI}\right)\sin\left(\frac{n\pi s}{L}\right) ds = 0. \tag{6.192}$$

This integral reduces to the following equation for the most important loading case, that in which the load is equal to the critical buckling load. This corresponds to $n = 1$ in Eq. (6.186), resulting in $P = P_{\text{crit}}$. The Galerkin orthogonality condition then becomes

$$-A\left[1 + \frac{1}{2}\left(\frac{\pi A}{2L}\right)^2 + \frac{3}{2}\left(\frac{\pi A}{2L}\right)^4 + \cdots\right] + \frac{PA}{P_{\text{crit}}} = 0. \tag{6.193}$$

By neglecting terms of A^4 and higher in the above equation and dropping the trivial solution $A = 0$, the amplitude at any arbitrary load is found to be

$$A = \frac{2L}{\sqrt{\pi}}\sqrt{\frac{P}{P_{\text{crit}}} - 1}. \tag{6.194}$$

This result has been found to agree extremely well for $A \leq L/5$.

EXAMPLE 6.17: HEAT CONDUCTION EXAMPLE

Consider the situation of an infinite wall of unit thickness, having a thermal conductivity of $k = k_0(1 + \alpha U)$, where U is the temperature at any position x within the wall at any time t. This is shown pictorially in Fig. 6.11. The governing differential equation for this case is

$$\rho c \frac{\partial U}{\partial t} = \frac{\partial}{\partial x}\left\{k \frac{\partial U}{\partial x}\right\}. \tag{6.195}$$

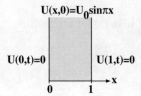

Figure 6.11. Temperature variations in an infinite wall.

If the time domain is transformed to a dimensionless variable, τ, where

$$\tau = \frac{k_0 t}{\rho c}, \tag{6.196}$$

then the governing differential equation becomes

$$\frac{\partial U}{\partial \tau} = (1 + \alpha U)\frac{\partial^2 U}{\partial \tau^2} + \alpha \left(\frac{\partial U}{\partial x}\right)^2. \tag{6.197}$$

This equation is subject to the boundary conditions

$$U(0, \tau) = U(1, \tau) = 0, \qquad U(x, 0) = U_0 \sin \pi x. \tag{6.198}$$

A Galerkin approximating solution function is assumed of the form

$$\tilde{U}(x, \tau) = \sum_{n=1}^{N} T_n(\tau) \sin n\pi x, \qquad n = 1, 3, \ldots, N, \tag{6.199}$$

so that the approximating function can closely represent the boundary conditions. The reader might consider more closely why only the odd terms have been included here. For this approximating function, applying the boundary conditions yields

$$\begin{aligned} U(0, x) &= U_0 \sin \pi x \\ &= T_1(0) \sin \pi x + T_3(0) \sin 3\pi x \\ &\quad + T_5(0) \sin 5\pi x + \cdots + T_N(0) \sin N\pi x. \end{aligned} \tag{6.200}$$

From the boundary condition, the coefficients are seen to be

$$T_1(0) = U_0,$$
$$T_m(0) = 0, \qquad m = 3, 5, \ldots,$$

and the Galerkin residue is therefore

$$\epsilon = \frac{\partial U}{\partial \tau} - (1 + \alpha U)\frac{\partial^2 U}{\partial \tau^2} - \alpha \left(\frac{\partial U}{\partial x}\right)^2. \tag{6.201}$$

One next integrates over the x domain using the Galerkin orthogonality conditions and obtains

$$\frac{dT_1}{d\tau} = -\pi^2 T_1$$
$$+ 4\alpha\pi \left[-\frac{1}{3}T_1^2 - \frac{9}{35}T_3^2 - \frac{25}{99}T_5^2 + \frac{2}{15}T_1 T_3 + \frac{2}{105}T_1 T_5 + \frac{10}{63}T_3 T_5 \right],$$

6.5. Galerkin's Method

TABLE 6.4. Galerkin vs Finite Difference Solutions to Conduction in an Infinite Wall

Pseudotime, τ	$\alpha = 0$		$\alpha = 1$		$\alpha = 10$	
	Finite diff. solution	Galerkin solution	Finite diff. solution	Galerkin solution	Finite diff. solution	Galerkin solution
0.00	1.000	1.000	1.000	1.000	1.000	1.000,
0.01	0.907	0.906	0.843	0.871	0.562	0.647
0.02	0.822	0.821	0.729	0.763	0.390	0.466
0.03	0.746	0.744	0.637	0.671	0.293	0.356
0.04	0.676	0.674	0.559	0.592	0.230	0.282
0.05	0.613	0.611	0.493	0.524	0.186	0.230
0.06	0.556	0.553	0.437	0.464	0.153	0.191
0.07	0.504	0.501	0.388	0.414	0.129	0.161
0.08	0.457	0.454	0.345	0.368	0.110	0.137
0.09	0.414	0.411	0.308	0.329	0.094	0.118

$$\frac{dT_3}{d\tau} = -9\pi^2 T_3$$
$$+ 36\alpha\pi \left[-\frac{1}{15}T_1^2 - \frac{1}{9}T_3^2 - \frac{25}{273}T_5^2 - \frac{18}{35}T_1T_3 + \frac{10}{63}T_1T_5 - \frac{18}{55}T_3T_5 \right],$$

$$\frac{dT_5}{d\tau} = -25\pi^2 T_5$$
$$+ 100\alpha\pi \left[-\frac{1}{105}T_1^2 - \frac{9}{55}T_3^2 - \frac{1}{15}T_5^2 + \frac{10}{63}T_1T_3 - \frac{50}{99}T_1T_5 - \frac{50}{273}T_3T_5 \right],$$

subject to the initial conditions $T_1(0) = U_0$, $T_3(0) = T_5(0) = 0$. These may be readily solved using many methods. For example, one might continue to employ Galerkin's method and assume an exponential approximation for the T_n functions. Table 6.4 shows the results of solving this problem using Galerkin's method and the method of finite differences. The comparison is only for the midplane solution, $x = 0.5$.

EXAMPLE 6.18: NONLINEAR SPRING-MASS PROBLEM

Consider the problem of two masses vibrating between fixed linear springs and interconnected by a cubic spring. This is shown in Fig. 6.12. The resulting differential equations are thus

$$m\frac{d^2x_1}{dt^2} + kx_1 + k(x_1 - x_2)^3 = 0,$$
$$m\frac{d^2x_2}{dt^2} + kx_2 + k(x_2 - x_1)^3 = 0.$$

Addition and subtraction yield the following two equations:

$$m\left(\frac{d^2x_1}{dt^2} + \frac{d^2x_2}{dt^2}\right) + k(x_1 + x_2) = 0,$$
$$m\left(\frac{d^2x_1}{dt^2} - \frac{d^2x_2}{dt^2}\right) + k(x_1 - x_2) + 2k(x_1 - x_2)^3 = 0.$$

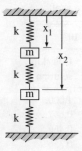

Figure 6.12. Two masses suspended on linear springs and connected by a cubic spring.

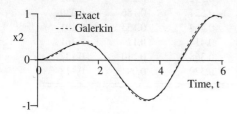

Figure 6.13. Galerkin versus exact solutions for two masses between nonlinear springs.

If two new displacement variables are defined as $z_1 = x_1 + x_2$ and $z_2 = x_1 - x_2$, two uncoupled differential equations are produced:

$$m\frac{d^2 z_1}{dt^2} + kz_1 = 0,$$

$$m\frac{d^2 z_2}{dt^2} + kz_2 + 2kz_2^3 = 0.$$

It is noteworthy that there are two modes of oscillation of this system:

1. Both masses oscillate together. The middle nonlinear spring is unimportant for this mode.
2. Both masses are displaced from one another initially and then allowed to oscillate and exchange energy.

Both of these modes yield sinusoidal modes. Figure 6.13 compares a numerical (labeled exact) and a Galerkin solution for the second mode. The solution for the displacement of mass 2 is portrayed, subject to the initial conditions $x_1(0) = 1$, $x_2(0) = 0$ and for $m = k = 1$. Note the growth in amplitude of x_2 as it absorbs energy from mass 1, while the total system energy remains constant. The Galerkin solution is accurate for this example to within approximately 3–4%. This error level is only for amplitudes and times of the order of 1. Because the Galerkin approximating function was optimized over a small time period, the errors grow at longer time periods because the periodicity of energy flow is much longer than the fundamental period of oscillation of the individual masses.

6.6 SUMMARY COMMENTS

The material presented in this chapter includes many powerful analytical techniques for obtaining approximate solutions to the full nonlinear equations for a system.

In Chapter 2, Section 2.6, a method for normalizing the governing equations was introduced. The benefits of this approach include identifying terms in the system model that may be neglected. In this manner, the complexity of the governing equations is reduced, thereby providing an opportunity to obtain an analytical solution to an approximate model. The methods presented in this chapter are very different; they allow approximate solutions to the full nonlinear model. Additionally, we have provided many approaches, allowing the modeler to attempt more than a single method, thereby gaining insight and confidence in the resulting solutions. The normalization method has an added advantage, however, in that it also may provide insight into the time and spatial scales of the problem. This added knowledge is important, for example, in the formulation of proper Galerkin approximating functions and limits on the integrals of the orthogonality conditions.

A key element in the methods presented in this chapter is the depth of insight the modeler brings to the solution attempt. This is generally true for most solution methods, if for no other reason than to allow the modeler to assess the validity of the solution. The modeler must always be looking for those apparently correct solutions, which, in fact, violate the *Principle of Minimum Astonishment*.[5]

6.7 PROBLEMS

The following problems apply approximate solution methods to basic nonlinear systems.

6.1 A spring-mass system is described by Eq. (6.202):

$$\frac{d^2x}{dt^2} + ax + bx^2 = A, \qquad x(0) = x_0, \quad \dot{x}(0) = 0. \tag{6.202}$$

Solve the homogeneous form of this equation ($A = 0$) by the method of perturbation. Start by assuming that

$$x = x_0 + bx_1,$$
$$\omega^2 = \omega_0^2 + b\omega_1^2,$$

where b is regarded as a "small" parameter. Use a symbolic algebra language to solve this problem.

6.2 The equation of motion for a mass supported on a cubic spring may be written as

$$\frac{d^2y}{dt^2} + a^2 y + b^2 y^3 = 0, \qquad y(0) = y_0, \quad \frac{dy}{dt}(0) = 0, \tag{6.203}$$

where $a, b = $ constants.

(1) Use the iteration technique to obtain a solution to this problem. Obtain a minimum of three iterations (y_0, y_1, y_2).
(2) Set $y(0) = y_0 = 1$, $a^2 = 10$, and $b^2 = 100$. Plot each of the iterative solutions covering a minimum time period of at least $\sqrt{a}t = 0 \cdots 2$.

You might wish to minimize the algebraic difficulties of the solution by first normalizing the governing equations.

---

[5] Andrew J. McPhate (ca. 1972): "No model or result should ever yield results which are counter to a well-developed base of intuition and experience."

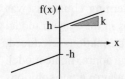

Figure 6.14. Functional relation of $f(x)$ to x.

6.3 The equation of motion for a metering pin damper is

$$\frac{d^2x}{dt^2} + \|x\|\frac{dx}{dt} + x = 0.$$

Use the method of slowly varying amplitude and phase to find a solution of the form

$$x = a(t)\sin(\omega t + \phi).$$

Assume that amplitude of motion is $x = a_0$ at time $t = 0$.

6.4 The equation of motion for an oscillatory system is given by the differential equation

$$\frac{d^2x}{dt^2} + f(x) = 0. \tag{6.204}$$

In this equation, $f(x)$ is a function represented by Fig. 6.14.

(1) Examine the system by the method of harmonic balance to find a solution of the form

$$x = a(t)\sin(\omega_0 t + \phi). \tag{6.205}$$

Assume that the amplitude of motion is $x = a_0$ at time $t = 0$.

(2) Attempt to verify your solution by at least two other methods for the special case of $k = 1$, $h = 5$, and $a_0 = 10$.

(3) Depict all of your results graphically and write a brief description of your observations and conclusions.

6.5 The equation of motion for a mass supported on a cubic spring is described in Problem 3.4.

(1) Find the equivalent linear system for this nonlinear system. Determine the amplitude, $a(t)$, and phase, $\phi(t)$.

(2) Set $y(0) = 1$, $a^2 = 10$, and $b^2 = 100$ and plot the solution using the methods of (a) harmonic balance, and (b) equivalent linear system.

(3) Discuss the solutions, especially considering the assumptions required in the method of harmonic balance.

6.6 A mass-spring system is described by the following equation:

$$\frac{d^2x}{dt^2} + x^2 = 0, \qquad x(0) = A, \qquad \frac{dx}{dt} = 0. \tag{6.206}$$

Convert this to an equivalent linear system and find the solution to this system. Also find the solution using the method of harmonic balance. If you use a symbolic algebra language to obtain a solution, include a script file of your work and any source files used.

6.7 A function $f(x)$ in the region $0 < x < 1$ is to be approximated by a linear combination of known functions, $\phi_i(x)$. Define the Galerkin residue, ϵ, as

$$\epsilon = f(x) - \sum C_i \phi_i(x), \tag{6.207}$$

6.7. Problems

where the C_i are constants. Show that, if the C_i are determined by applying the Galerkin procedure to ϵ, the integral of ϵ^2 over the range $0 < x < 1$ is minimized.

6.8 By use of Galerkin's method, obtain an approximate solution to the differential equation

$$\frac{d^2x}{dt^2} + 2.25x + [x - 1.5\sin(t)]^3 = 2\sin(t). \tag{6.208}$$

Assume a solution of the form

$$x = A_1 \sin(t) + A_3 \sin(3t). \tag{6.209}$$

6.9 The equation of motion for a metering pin damper is given in Problem 6.3. Using the approximation function

$$x = c_1 + c_2 e^{-kt}, \tag{6.210}$$

obtain the constants c_1, c_2, and k. For the special case of $v_0 = 1$, plot the exact and approximation functions and the difference between the two results over the time from $t = 0 \cdots 8$.

6.10 A spring-dashpot system is described by the following equation:

$$\frac{dx}{dt} + ax + bx^2 = 0, \qquad x(0) = X_0.$$

Solve this equation by Galerkin's method, assuming that the solution is of the form

$$x = Ce^{-kt}.$$

Find analytic expressions for the constants C and k. As a special case, set $X_0 = 1$, $a = 2$, $b = 1$ and find $x(t)$. Compare your results with the exact solution. Also, turn in a computer-generated plot of x versus t over the region $t[0, 2.5]$ showing both the Galerkin and the exact solutions.

CHAPTER 7

Stability of Nonlinear Systems

No discussion of solutions to differential equations would be complete without a presentation of methods to assess the stability of the model system. Since there is a great deal of information available on the stability of linear systems, this chapter focuses principally on nonlinear systems. It is essential that the reader understand that what is being discussed here relates to the stability of the model, or the system of equations. The system being studied will be stable only if the mathematical model contains sufficient detail to include the same stability values as the original system. If the system is unstable, then a complete model would have governing equations in which there were no bounded solutions. Mathematically, it may be possible to find a bounded solution to the equations of motion, but a small perturbation would result in a motion away from this solution.

The first critical issue in our discussion is a definition of stability. A linear system is considered to be stable if, for every bounded input, a bounded output results. This is a difficult definition to apply to a nonlinear system because the relationship between input and output is nonlinear and may contain surprises, such as spurious frequency components not equal to any forcing frequency or natural frequency of the system, limit cycles unrelated to the input, or regional instability. Another definition used for the stability of linear systems is that such a system is declared stable if its unforced response decays asymptotically to zero. For example, the unforced transient response of any linear system may be represented by

$$x(t) = A_1 e^{\lambda_1 t} + A_2 e^{\lambda_2 t} + \cdots + A_n e^{\lambda_n t}, \tag{7.1}$$

where λ_i denotes the roots of the characteristic equations. Stability exists when the real part of λ_i is negative (Routh test, Nyquist test). The roots for linear systems are dependent *only* on the characteristics of the system, not on the input forcing function or the boundary conditions.

A tentative or trial definition for the stability of the model of a nonlinear system is that it is stable if any bounded input produces an output that is bounded. A second definition is that such a system is stable if its transient response decays to zero. Two simple examples of nonlinear spring-dashpot systems having very different solutions and stability considerations are presented next.

EXAMPLE 7.1: FORCED NONLINEAR SPRING-DASHPOT SYSTEM

A nonlinear spring and dashpot system with a constant force has a differential equation given as

$$\frac{dx}{dt} + x^2 = A, \qquad x(0) = 0, \tag{7.2}$$

where A is a constant. The solution is found to be

$$x = \sqrt{A} \tanh \sqrt{A}\, t, \qquad A > 0, \tag{7.3}$$

$$x = -\sqrt{-A} \tan \sqrt{-A}\, t, \qquad A < 0, \tag{7.4}$$

which yields the conclusion that the model system is stable for $A > 0$ but unstable for $A < 0$.

EXAMPLE 7.2: UNFORCED NONLINEAR SPRING-DASHPOT SYSTEM

A nonlinear spring and dashpot system has a differential equation given as

$$\frac{dx}{dt} + x^2 = 0, \qquad x(0) = x_0, \tag{7.5}$$

which has a solution

$$x(t) = \frac{x_0}{x_0 t + 1}. \tag{7.6}$$

The stability of this system model is assured for $x_0 > 0$ but it becomes unstable for $x_0 < 0$.

To formulate a general approach to the analysis of the stability of system models, the problem may be subdivided into the following classes:

Class I. Homogeneous systems with small initial conditions;
Class II. Homogeneous systems with large initial conditions;
Class III. Systems with small forcing functions, homogeneous initial conditions;
Class IV. Systems with large forcing functions, homogeneous initial conditions.

For the first two classes, some successes may be noted, but the second two classes of problems are much more difficult. The stability of nonlinear systems remains a research area.

7.1 ROUTH METHOD

A complete description of the Routh method is given by Chestnut and Mayer [14] but it is summarized here for completeness. This method is a technique for investigating the stability of linear systems of the form

$$a_n \frac{d^n x}{dt^n} + a_{n-1} \frac{d^{n-1} x}{dt^{n-1}} + \cdots + a_2 \frac{d^2 x}{dt^2} + a_1 \frac{dx}{dt} + a_0 x = 0, \tag{7.7}$$

which has solutions of the form $x = x_0 e^{\lambda t}$. The characteristic equation for such systems to determine the eigenvalues, λ, that yield acceptable solutions may be written as

$$a_n \lambda^n + a_{n-1} \lambda^{n-1} + \cdots + a_1 \lambda + a_0 = 0. \tag{7.8}$$

The Routh array is given in terms of the coefficients, a_i, of Eq. (7.8):

$$\begin{vmatrix} a_n & a_{n-2} & a_{n-4} \\ a_{n-1} & a_{n-3} & \\ b_{n-1} & b_{n-3} & \\ c_{n-1} & c_{n-3} & \\ \vdots & & \end{vmatrix}. \qquad (7.9)$$

You must generate as many terms in the first column as $n+1$ terms of the nth-order equation. The b and c terms are given by

$$b_{n-1} = \frac{a_{n-1}a_{n-2} - a_n a_{n-3}}{a_{n-1}}, \qquad (7.10)$$

$$b_{n-3} = \frac{a_{n-1}a_{n-4} - a_n a_{n-5}}{a_{n-1}}, \qquad (7.11)$$

$$c_{n-1} = \frac{b_{n-1}b_{n-3} - a_{n-1}b_{n-3}}{b_{n-1}}. \qquad (7.12)$$

The number of sign changes in the first column yields the number of positive (unstable) characteristic roots, λ. If there are no positive roots for λ, then the governing equation is considered stable.

EXAMPLE 7.3: FLYBALL GOVERNOR

The governing equations of motion for a flyball governor were derived previously in Chapter 3, Example 3.5, resulting in the equations

$$m\frac{d^2 y}{dt^2} = m\omega_0^2(\cos^2\phi_0 - \sin^2\phi_0)y + 2mx\omega_0 \sin\phi_0 \cos\phi_0 - mgy \cos\phi_0 - b\frac{dy}{dt}, \qquad (7.13)$$

$$I\frac{dx}{dt} = -ky \sin\phi_0. \qquad (7.14)$$

A single third-order equation may be obtained from these two equations:

$$m\frac{d^3 y}{dt^3} + b\frac{d^2 y}{dt^2} + \left(\frac{mg \sin^2\phi_0}{\cos\phi_0}\right)\frac{dy}{dt} + \left(\frac{2mgk \sin^2\phi_0}{I\omega_0}\right)y = 0. \qquad (7.15)$$

The characteristic equation is thus

$$m\lambda^3 + b\lambda^2 + \frac{mg \sin^2\phi_0}{\cos\phi_0}\lambda + \frac{2mgk \sin^2\phi_0}{I\omega_0} = 0. \qquad (7.16)$$

Because the governing equation is linear, stability tests can be applied in the form of the Nyquist or Routh-Hurwitz method. Applying the Routh method, the Routh array defined in Eq. (7.9) becomes

$$\begin{vmatrix} m & \frac{mg \sin^2\phi_0}{\cos\phi_0} \\ b & \frac{2mkg \sin^2\phi_0}{I\omega_0} \\ \xi & 0 \\ \frac{2mgk \sin^2\phi_0}{I\omega_0} & 0 \end{vmatrix}, \qquad (7.17)$$

where

$$\xi = \frac{mg \sin^2 \phi_0}{\cos \phi_0} - \frac{2m^2 kg \sin^2 \phi_0}{bI\omega_0}. \tag{7.18}$$

All of the terms in column 1 of the Routh array are positive except ξ, which can switch signs. The number of positive roots is the number of unstable regions of the solution. This number is equal to the number of sign changes in this first column. The stability criterion for the system is thus

$$\frac{mg \sin^2 \phi_0}{\cos \phi_0} - \frac{2m^2 kg \sin^2 \phi_0}{bI\omega_0} > 0 \tag{7.19}$$

or

$$mg \sin^2 \phi_0 \left[\frac{1}{\cos \phi_0} - \frac{2mk}{bI\omega_0} \right] > 0, \tag{7.20}$$

which simplifies to

$$\frac{bI}{m} \geq \frac{2k \cos \phi_0}{\omega_0}. \tag{7.21}$$

This becomes the stability criterion for the system, which may be rewritten as

$$\frac{2km \cos \phi_0}{bI\omega_0} < 1. \tag{7.22}$$

The stability of the solution for the linearized form of the flyball governor problem therefore is limited to a small region near the operating point. Note also that the stability is a function only of the characteristics of the system, not the forcing function.

7.2 SINGULAR-POINT ANALYSIS

Understanding the stability of nonlinear systems can, under certain conditions, be enhanced by examining the behavior of such systems in the region immediate to their singular points. Singularities are points of equilibrium, but they also may be points at which the system is unconditionally stable or unstable. A ball on the top of a hill is an example of such quasistable equilibrium. Any small perturbation of the ball from its equilibrium position results in a motion that rapidly diverges from the ball's former equilibrium position.

The qualitative insight afforded by examining the behavior of systems at their singularities is considered an important tool. The method is both analytical and graphical, thereby taking advantage of the analytical and visual powers of comprehension inherent in the human brain. Unlike with the Routh method, we are not restricting this discussion to linear or linearized systems. The singular point stability analysis, however, is restricted to the region immediate to the equilibrium points.

7.2.1 General System

Consider an autonomous system described by the following differential equations:

$$\frac{dx}{dt} = P(x, y), \tag{7.23}$$

$$\frac{dy}{dt} = Q(x, y), \tag{7.24}$$

where P and Q may be nonlinear and are functions of x, y. Division of Eq. (7.24) by Eq. (7.23) yields

$$\frac{dy}{dx} = \frac{Q(x, y)}{P(x, y)}. \tag{7.25}$$

The form of this equation eliminates the independent variable, t, from the system and *limits the analysis to cases in which the forcing function is absent or extremely simple.*

A new term is introduced here to define points in the function space of a system where it becomes stable. A *singular point* is defined as one where

$$P(x_s, y_s) \equiv Q(x_s, y_s) \equiv 0 \text{ or } \infty. \tag{7.26}$$

Such points also are seen to be points of equilibrium since $dx/dt = 0$ and $dy/dt = 0$. To begin an analysis of the stability of a system in the region near a singular point, two new variables are introduced:

$$x = x_s + u, \tag{7.27}$$
$$y = y_s + v, \tag{7.28}$$

where x_s, y_s are the constants representing the coordinates of the singular point. Thus, Eqs. (7.27) and (7.28) are nothing more than a translation of the singularity to the origin of the u, v coordinate system as shown in Fig. 7.1. The resulting two new variables, u and v, contain the variation of the behavior of the system in a region immediate to the singularity. If these definitions are substituted into Eq. (7.25) and P, Q are expanded in a Taylor series about the singular point, x_s, y_s, then

$$\frac{dy}{dx} = \frac{dv}{du} = \frac{Q(x_s, y_s) + cu + dv + [c_2 u^2 + d_2 v^2 + f_2 uv + \cdots]}{P(x_s, y_s) + au + bv + [a_2 u^2 + b_2 v^2 + e_2 uv + \cdots]}, \tag{7.29}$$

but $Q(x_s, y_s) = P(x_s, y_s) = 0$ and if terms containing only the higher-order terms $(u^2, v^2, uv, \ldots)$ are collected into two new variables, P_2, Q_2,

$$P_2 = [a_2 u^2 + b_2 v^2 + e_2 uv + \cdots], \tag{7.30}$$
$$Q_2 = [c_2 u^2 + d_2 v^2 + f_2 uv + \cdots], \tag{7.31}$$

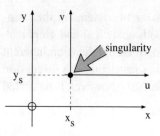

Figure 7.1. Transformation of origin to singular point.

7.2. Singular-Point Analysis

this yields

$$\frac{dv}{du} = \frac{cu + dv + Q_2(u, v)}{au + bv + P_2(u, v)}. \tag{7.32}$$

Poincaré [33],[1] working on the local behavior of differential equations describing celestial motion, found that in the region of singularities he could reduce the complexity of consideration of stability. Since P_2 and Q_2 in Eq. (7.32) contain only higher-order terms, the stability of the system in the neighborhood of the singular point (x_s, y_s) may be determined by examining only

$$\frac{dv}{du} = \frac{cu + dv}{au + bv} \tag{7.33}$$

under the constraint that $bc \neq ad$. If the higher-power terms, (P_2, Q_2), in Eq. (7.32) are the only terms and there are no first-order terms, then the singularity is not classified as a simple singularity. This section is directed only to systems having "simple" singularities.

7.2.2 Linear Transformations

The transformation represented by Eqs. (7.27) and (7.28) amounts to a simple shift in the origin to the singular point. Thus, the derivatives in this region for u, v are approximated by

$$\frac{du}{dt} = au + bv, \tag{7.34}$$

$$\frac{dv}{dt} = cu + dv. \tag{7.35}$$

Substituting $x_1 = u$, $x_2 = v$, $a_{11} = a$, $a_{12} = b$, $a_{21} = c$, and $a_{22} = d$ into the above equations yields

$$\dot{x}_1 = a_{11}x_1 + a_{12}x_2, \tag{7.36}$$

$$\dot{x}_2 = a_{21}x_1 + a_{22}x_2, \tag{7.37}$$

which also may be represented as a matrix of equations:

$$\{\dot{x}\} = [A]\{x\}. \tag{7.38}$$

The solution of Eq. (7.38) is

$$\{x\} = e^{\lambda t}\{C\}, \tag{7.39}$$

where

$$\{C\} = \begin{Bmatrix} C_1 \\ C_2 \end{Bmatrix}. \tag{7.40}$$

By differentiating Eq. (7.39) and inserting this into Eq. (7.38), the following is obtained:

$$\{\dot{x}\} = \lambda e^{\lambda t}\{C\} = \lambda\{x\} = [A]\{x\}. \tag{7.41}$$

---

[1] Henri Poincaré (1854–1912).

The identity matrix, I, is defined as a matrix having unit value for all diagonal elements and off-diagonal values of zero:

$$I = \begin{bmatrix} 1 & 0 & 0 & 0 \\ 0 & 1 & 0 & 0 \\ 0 & 0 & 1 & 0 \\ 0 & 0 & 0 & 1 \end{bmatrix}. \tag{7.42}$$

It can be shown that

$$[A]\{x\} = \lambda I\{x\}. \tag{7.43}$$

Introducing this into Eq. (7.41) yields

$$\|[A] - [\lambda I]\|\{x\} = \{Z\}, \tag{7.44}$$

where Z is the zero matrix containing all zeroes and the row-column order of the equation. This system of equations has a nonzero solution only if the determinant is also zero, or

$$\|[A] - [\lambda I]\| = 0 \tag{7.45}$$

or

$$\left\| \begin{bmatrix} a_{11} & a_{12} \\ a_{21} & a_{22} \end{bmatrix} - \begin{bmatrix} \lambda & 0 \\ 0 & \lambda \end{bmatrix} \right\| = 0 \tag{7.46}$$

or

$$\left\| \begin{bmatrix} (a_{11} - \lambda) & a_{12} \\ a_{21} & (a_{22} - \lambda) \end{bmatrix} \right\| = 0. \tag{7.47}$$

These determinants represent the characteristic equations for the system. Equations (7.38) are, in general, a coupled set of equations since both x_1 and x_2 appear in both differential equations. They are only uncoupled for the special case of $a_{12} = a_{21} = 0$.

EXAMPLE 7.4: SPRING-MASS-DASHPOT STABILITY

Consider the following set of differential equations describing the forced motion of a spring-mass-dashpot system:

$$\frac{dy}{dt} = v, \tag{7.48}$$

$$\frac{dv}{dt} = \frac{F}{m} - \frac{c}{m}v - \frac{k}{m}y. \tag{7.49}$$

This yields the ratio of equations

$$\frac{dv}{dy} = \left(\frac{F}{m} - \frac{c}{m}v - \frac{k}{m}y\right) \Big/ v. \tag{7.50}$$

7.2. Singular-Point Analysis

If we define the translation of origin to the singularity (v_s, y_s), or $v = v_s + q$, $y = y_s + r$, this becomes

$$\frac{dv}{dy} = \frac{\left(\frac{F}{m} - \frac{c}{m}v - \frac{k}{m}y\right) - \frac{c}{m}q - \frac{k}{m}r}{v_s + q}. \tag{7.51}$$

If we take the singularity $v_s = 0$, $y_s = F/k$, and drop the higher-order terms, then Eq. (7.51) becomes

$$\frac{dq}{dr} = \frac{-\left(\frac{c}{m}q - \frac{k}{m}r\right)}{q}. \tag{7.52}$$

or

$$\begin{Bmatrix} \dot{q} \\ \dot{r} \end{Bmatrix} = [A] \begin{Bmatrix} q \\ r \end{Bmatrix} = \begin{bmatrix} -c/m & -k/m \\ 0 & 1 \end{bmatrix} \begin{Bmatrix} q \\ r \end{Bmatrix}. \tag{7.53}$$

This yields solutions of the form

$$\begin{Bmatrix} q \\ r \end{Bmatrix} = e^{\lambda t} \{C\}. \tag{7.54}$$

Inserting this into the differential equation yields a characteristic equation analagous to Eq. (7.47):

$$\begin{bmatrix} -c/m & -k/m \\ 0 & 1 \end{bmatrix} - \begin{bmatrix} \lambda & 0 \\ 0 & \lambda \end{bmatrix} = 0. \tag{7.55}$$

This can be shown to yield the characteristic roots $\lambda = 1, -c/m$. The first root is unstable because it leads to a positive exponential solution. The second root is stable.

The effects of a linear transformation of the governing differential equations are considered next. If the transformation is linear, then the characteristic equations, their roots, and the resulting stability of the systems must all be identical. Consider the following general linear transformation from our original variables (x_1, x_2) to a new set (y_1, y_2):

$$x_1 = P_{11} y_1 + P_{12} y_2, \tag{7.56}$$
$$x_2 = P_{21} y_1 + P_{22} y_2, \tag{7.57}$$

where P_{ij} are constants. The transformation results in a relative rotation, stretching, and/or rotation of the system in the same manner as the simple geometry depicted in Fig. 7.2. The variables y_1, y_2 are the new independent variables. In matrix form, this may be written as

$$\{x\} = [P]\{y\} = \begin{bmatrix} P_{11} & P_{12} \\ P_{21} & P_{22} \end{bmatrix} \begin{Bmatrix} y_1 \\ y_2 \end{Bmatrix}. \tag{7.58}$$

If $[P]$ has an inverse, $[P]^{-1}$, then it is possible to write

$$[P][P]^{-1} = [I],$$
$$[P]^{-1}[P] = [I], \tag{7.59}$$

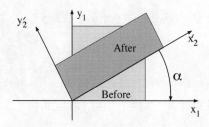

Figure 7.2. Stretching, compression, rotation from real-valued, linear transformations.

which may be used with Eq. (7.58) to obtain

$$[P]^{-1}\{x\} = [P]^{-1}[P]\{y\}$$
$$= \{y\}. \tag{7.60}$$

Combining Eqs. (7.38), (7.58), and (7.59) yields

$$\{\dot{y}\} = [B]\{y\}, \tag{7.61}$$

where

$$[B] = [P]^{-1}[A][P]. \tag{7.62}$$

By analogy with previous results, the characteristic equation for this system is

$$\{[B] - \lambda[I]\}. \tag{7.63}$$

Now,

$$[B] - \lambda_B[I] = [P]^{-1}[A][P] - \lambda_B[I],$$
$$[B] - \lambda_B[I] = [P]^{-1}\{[A] - \lambda_B[I]\}[P]. \tag{7.64}$$

The roots to this system may be calculated as

$$\begin{vmatrix} (a - \lambda_B) & b \\ c & (d - \lambda_B) \end{vmatrix} = 0. \tag{7.65}$$

This equation is identical to the original characteristic equation. When expanded, it becomes

$$\lambda^2 - (a+d)\lambda - (bc - ad) = 0, \tag{7.66}$$

where $a = a_{11}, b = a_{12}, c = a_{21}$, and $d = a_{22}$.

7.2.3 Classification of Singularities

The types of singularities resulting from a second-order system can be classified by examining the roots that result from the solution of Eq. (7.66). This classification is shown in Table 7.1. These roots may be represented as

$$\lambda_{1,2} = \tfrac{1}{2}\Big[(a+d) \pm \sqrt{(a+d)^2 + 4(bc - ad)}\,\Big]. \tag{7.67}$$

7.2. Singular-Point Analysis

TABLE 7.1. Classification of the Four Types of Singularities

Resultant root	Type
Real with same sign	NODE
Real with opposite sign	SADDLE
Pure imaginary	CENTER
Complex conjugate	FOCUS

The equations in normal form are thus

$$\dot{y}_1 = \lambda_1 y_1, \tag{7.68}$$
$$\dot{y}_2 = \lambda_2 y_2. \tag{7.69}$$

The solutions for λ are thus either real or complex based on the sign of the radical term $r = (a+d)^2 + 4(bc - ad)$. Each of the possible root classifications will be discussed in the following sections.

7.2.3.1 Real Roots, Same Sign

Roots that fall into this category result from $[(a+d)^2 + 4(bc - ad)] > 0$. The ratio of differential equations yields

$$\frac{\dot{y}_2}{\dot{y}_1} = \frac{dy_2}{dy_1} = \frac{\lambda_2 y_2}{\lambda_1 y_1}. \tag{7.70}$$

If both roots have the same signs, then

$$\frac{\lambda_2}{\lambda_1} \doteq A > 0; \tag{7.71}$$

thus

$$\frac{dy_2}{dy_1} = A \frac{y_2}{y_1}, \tag{7.72}$$

which may be integrated to obtain

$$y_2 = y_2(0) \left[\frac{y_1}{y_1(0)}\right]^A, \tag{7.73}$$

where $y_1(0)$ and $y_2(0)$ are the initial conditions. The three possibilities for A are shown in Fig. 7.3. The governing equations are

$$\dot{y}_2 = \lambda_2 y_2, \tag{7.74}$$
$$\dot{y}_1 = \lambda_1 y_1. \tag{7.75}$$

If λ_1, λ_2 are positive, the solution diverges from the origin. This is classified as *unstable*. Negative values for λ_1, λ_2 result in *stable* solutions.

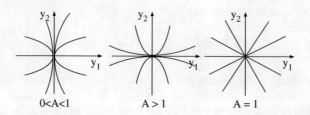

Figure 7.3. Graphical types yielding nodes.

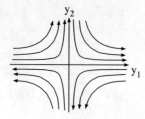

Figure 7.4. Real roots with different signs produce SADDLEs.

7.2.3.2 Real Roots, Different Sign

Roots that fall into this category result from $[(a+d)^2 + 4(bc-ad)] > 0$. The ratio of differential equations yields

$$\frac{\dot{y}_2}{\dot{y}_1} = \frac{dy_2}{dy_1} = \frac{\lambda_2 y_2}{\lambda_1 y_1}. \tag{7.76}$$

If both roots have the opposite sign, then

$$\frac{\lambda_2}{\lambda_1} \doteq -B < 0; \tag{7.77}$$

thus, applying these roots to Eqs. (7.74) and (7.75), we obtain

$$y_2 y_1^B = y_2(0) y_1^B(0). \tag{7.78}$$

This is depicted graphically in Fig. 7.4. The nature of this solution is that it is *always unstable*. The resulting graphical appearance of the solution in the vicinity of the singularity classifies this as a SADDLE point.

7.2.3.3 Complex Roots (Pure Imaginary)

In this section we consider the roots of the characteristic equation of the form

$$\lambda_{1,2} = -\alpha \pm j\beta. \tag{7.79}$$

This results in

$$\dot{y}_2 = y_2(-\alpha - j\beta), \tag{7.80}$$
$$\dot{y}_1 = y_1(-\alpha + j\beta). \tag{7.81}$$

An equivalent system is

$$\dot{y}_1 = y_2, \tag{7.82}$$
$$\dot{y}_2 = -y_1(\alpha^2 + \beta^2) - 2\alpha y_2. \tag{7.83}$$

7.2. Singular-Point Analysis

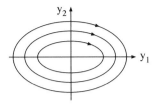

Figure 7.5. Complex pure imaginary roots produce CENTERs.

From this, we obtain

$$\frac{dy_2}{dy_1} = -\frac{(\alpha^2 + \beta^2)y_1 + 2\alpha y_2}{y_2}. \tag{7.84}$$

If the roots are pure imaginary, $\alpha = 0$, and

$$\frac{dy_2}{dy_1} = -\frac{\beta^2 y_1}{y_2}, \tag{7.85}$$

which can be integrated to obtain

$$y_2^2 + \beta^2 y_1^2 = y_2^2(0) + \beta^2 y_1^2(0). \tag{7.86}$$

This is depicted graphically in Fig. 7.5. The nature of this solution is that it is always stable in the sense that limit cycles exhibit bounded behavior. It is termed Lyapunov stable and, because of its appearance in the phase plane, it is defined as a CENTER.

7.2.3.4 Complex Conjugate Roots

This development proceeds as before, but the integration of Eq. (7.84) yields

$$y_2^2 + 2\alpha y_1 y_2 + (\alpha^2 + \beta^2)y_1^2 = C^2 \exp\left[\frac{2\alpha}{\beta} \arctan \frac{y_2 + \alpha y_1}{\beta y_1}\right], \tag{7.87}$$

where C^2 depends on the initial conditions. If this equation is transformed using

$$X = \beta y_1, \tag{7.88}$$
$$Y = y_2 + \alpha y_1, \tag{7.89}$$

for $\beta > 0$ we obtain

$$X^2 + Y^2 = C^2 \exp\left[\frac{2\alpha}{\beta} \arctan \frac{Y}{X}\right]. \tag{7.90}$$

We now set $X = \rho \cos\phi$ and $Y = \rho \sin\phi$, which can be interpreted graphically as shown in Fig. 7.6. Applying this transformation, we obtain

$$\rho = C \exp\left[\frac{\alpha\phi}{\beta}\right]. \tag{7.91}$$

The resulting solution depends on the sign of α and is shown graphically in Fig. 7.7. This behavior is typed because of its graphical appearance in the phase plane as a FOCUS. The associated singularity is called a FOCUS point.

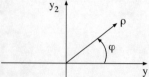

Figure 7.6. Definition of transformation.

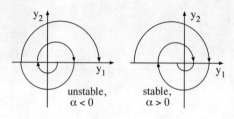

Figure 7.7. Complex conjugate roots produce a FOCUS.

In summary, the roots of the system are found as

$$\lambda_{1,2} = \tfrac{1}{2}\left[(a+d) \pm \sqrt{(a+d)^2 + 4(bc - ad)}\right]. \tag{7.92}$$

If $(a+d)^2 = 4(bc - ad) > 0$, then the result is a NODE or SADDLE.

7.2.4 Examples of Singularity Analysis

In the following examples, several physical systems that were presented earlier are examined with respect to their stability and then are classified.

EXAMPLE 7.5: DIRECTIONAL CONTROL OF VEHICLES

For small motions, the heading angle of a car, plane, or ship is given by the differential equation

$$\ddot{\theta} + k_1\dot{\theta} - k_2\theta = 0, \quad (k_1, k_2) > 0. \tag{7.93}$$

Introducing the velocity as $y = \dot{\theta}$, the differential equation becomes

$$\frac{dy}{d\theta} = \frac{k_2\theta - k_1 y}{y}. \tag{7.94}$$

Unforced linear systems such as this can be shown to have only one singular point which is located at the origin ($\theta_s = 0$, $y_s = 0$). In Eq. (7.94), the only point at which the denominator and numerator vanish is at the origin.

The problem is first translated to the singular point using the transformation

$$y = y_s + v,$$
$$\theta = \theta_s + u.$$

7.2. Singular-Point Analysis

In this instance, this is a meaningless variable change since the singularity is already at the origin. This transformation becomes

$$\frac{dy}{d\theta} = \frac{dv}{du} = \frac{k_2 u - k_1 v}{v} = \frac{cu + dv}{au + bv}. \tag{7.95}$$

From this equation, the canonical coefficients are found to be $a = 0$, $b = 1$, $c = k_2$, and $d = -k_1$. These can be inserted into the radical term in Eq. (7.67) to yield

$$r = (a + d)^2 + 4(bc - ad) = k_1^2 + 4k_2. \tag{7.96}$$

In this instance, $r > 0$ since both k_1 and k_2 are positive. Therefore, the solutions to the differential equation may be only of the form NODE or SADDLE as determined by the term $(bc - ad)$. For this example, $(bc - ad) = k_2 > 0$ and the solutions are always of the form SADDLE. Had $(bc - ad)$ been negative, the solution would have been classified as a NODE.

Because the form of the solution is a SADDLE, it is an unstable system. Additional information about the solution may be derived from the slope of the solutions at the origin as given in Eq. (7.95). This must be evaluated using L'Hôpital's rule. This rule states that, for two functions f and g that can be differentiated at every location except perhaps at $x = x_0$ within an interval (x_1, x_2), if $g'(x) \neq 0$ for all $x \neq x_0$ and if the ratio $f(x_0)/g(x_0)$ is indeterminate, it can be shown [48] that

$$\lim_{x \to x_0} \frac{f(x)}{g(x)} = \lim_{x \to x_0} \frac{f'(x)}{g'(x)}. \tag{7.97}$$

Using this rule, the slope at the origin, s, then can be written as

$$s = \left.\frac{dv}{du}\right|_{(0,0)} = \left.\frac{k_2 - k_1 \frac{dv}{du}}{\frac{dv}{du}}\right|_{(0,0)} \frac{k_2 - k_1 s}{s}, \tag{7.98}$$

which simplifies to the quadratic equation for s,

$$s^2 + k_1 s - k_2 = 0, \tag{7.99}$$

having the solution

$$s = \frac{-k_1 \pm \sqrt{k_1^2 + 4k_2}}{2}. \tag{7.100}$$

The solution to this problem is found using the characteristic roots, λ_1, λ_2, found from the characteristic equation

$$\lambda_{1,2} = \frac{-k_1 \pm \sqrt{k_1^2 + 4k_2}}{2}. \tag{7.101}$$

The obvious equivalence between Eqs. (7.100) and (7.101) is not by chance. The solution obtained using $\lambda_{1,2}$ is

$$\theta(t) = c_1 e^{\lambda_1 t} + c_2 e^{\lambda_2 t}, \tag{7.102}$$

$$y(t) = c_1 \lambda_1 e^{\lambda_1 t} + c_2 \lambda_2 e^{\lambda_2 t}, \tag{7.103}$$

where c_1, c_2 are arbitrary constants. The two cases of $c_1 = 0$ and $c_2 = 0$ yield the phase plane solutions

$$y = \lambda_1 \theta \quad (c_1 = 0), \tag{7.104}$$
$$y = \lambda_2 \theta \quad (c_2 = 0). \tag{7.105}$$

These are two straight lines having slopes of $\lambda_{1,2}$ through the origin.

EXAMPLE 7.6: MASS ON A NONLINEAR SPRING

The differential equation for a system with a fifth-order nonlinear spring is

$$\frac{d^2 x}{dt^2} + 4x - 5x^3 + x^5 = 0, \tag{7.106}$$

which results in

$$\frac{dy}{dx} = \frac{-x(x^4 - 5x^2 + 4)}{y} = -\frac{x(x^2 - 1)(x^2 - 4)}{y}. \tag{7.107}$$

The denominator and numerator vanish simultaneously at the five singular points $(x_s, y_s) = (0, 0), (\pm 1, 0)$, and $(\pm 2, 0)$. Each of these singularities must be examined to determine its behavior.

Singular point at (0, 0). The transformation is defined as

$$x = 0 + u,$$
$$y = 0 + v.$$

This produces the equation

$$\left.\frac{dv}{du}\right|_{(0,0)} = \frac{-4u + (5u^3 - u^5)}{v}. \tag{7.108}$$

Poincaré proved that the higher powers of u, v could be dropped with no change in the resulting stability. This action greatly simplifies the analysis and results in

$$\left.\frac{dv}{du}\right|_{(0,0)} = -\frac{4u}{v} = \frac{cu + dv}{au + bv}, \tag{7.109}$$

which yields $a = 0$, $b = 1$, $c = -4$, and $d = 0$. The radical, r, is examined next:

$$r = (a + d)^2 + 4(bc - ad) = -16. \tag{7.110}$$

Because $r < 0$, this singularity is either a FOCUS or a CENTER. Since $(a+d) = 0$, the singular point at the origin can be classified as a CENTER.

Singular point at (1, 0). The transformation is defined as

$$x = 1 + u,$$
$$y = 0 + v.$$

7.2. Singular-Point Analysis

This produces the equation

$$\left.\frac{dv}{du}\right|_{(1,0)} = \frac{6u + (5u^2 - 5u^3 - 5u^4 - u^5)}{v}. \tag{7.111}$$

Dropping the higher powers of u, v results in

$$\left.\frac{dv}{du}\right|_{(0,0)} = \frac{6u}{v} = \frac{cu + dv}{au + bv}, \tag{7.112}$$

which yields $a = 0$, $b = 1$, $c = 6$, and $d = 0$. The radical term in the characteristic equation, r, is examined next:

$$r = (a + d)^2 + 4(bc - ad) = 24. \tag{7.113}$$

Because $r > 0$, this singularity is either a NODE or a SADDLE. Since $(bc - ad) = 24 > 0$, the singular point at $(1, 0)$ can be classified as a SADDLE.

Singular point at $(2, 0)$. The transformation is defined as

$$x = 2 + u,$$
$$y = 0 + v.$$

This produces the equation

$$\left.\frac{dv}{du}\right|_{(1,0)} = -\frac{24u + \text{HOT}}{v}. \tag{7.114}$$

Dropping the higher powers of u, v results in

$$\left.\frac{dv}{du}\right|_{(0,0)} = \frac{-24u}{v} = \frac{cu + dv}{au + bv}, \tag{7.115}$$

which yields $a = 0$, $b = 1$, $c = -24$, and $d = 0$. The radical, r is determined to be

$$r = (a + d)^2 + 4(bc - ad) = -96. \tag{7.116}$$

Because $r < 0$, this singularity is either a FOCUS or a CENTER. Since $(a+d) = 0$, the singular point at $(2, 0)$ can be classified as a CENTER.

Singular points at $(-1, 0)$, $(-2, 0)$. It is readily shown that by symmetry we have SADDLE singularities at $(\pm 1, 0)$ and CENTER singularities at $(\pm 2, 0)$.

The phase plane now can be constructed in a general manner as shown in Fig. 7.8. The approximate behavior of each type of singularity is sketched in the appropriate location

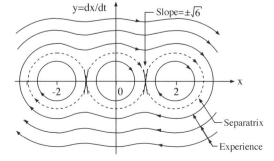

Figure 7.8. Singular point stability analysis for a mass on a fifth-order spring.

as shown in **A**. Based on experience, the behavior of the system over the entire phase plane can be sketched as in **B**.

EXAMPLE 7.7: INTERACTING NONLINEAR SYSTEMS

Consider the following differential equations for two interacting nonlinear systems:

$$\frac{dx}{dt} = k_1 x - k_3 xy, \tag{7.117}$$

$$\frac{dy}{dt} = k_2 y - k_4 xy, \tag{7.118}$$

or

$$\frac{dy}{dx} = \frac{k_2 y - k_4 xy}{k_1 x - k_3 xy} = \frac{y(k_2 - k_4 x)}{x(k_1 - k_3 y)}. \tag{7.119}$$

This system has two singular points located at $(0, 0)$ and $(k_2/k_4, k_1/k_3)$. The behavior of the system may be examined at the regions near these two singularities.

Singular point at (0, 0). The transformation to the origin is

$$x = 0 + u,$$
$$y = 0 + v,$$

which produces the equation

$$\left.\frac{dv}{du}\right|_{(0,0)} = \frac{k_2 v}{k_1 u} = \frac{cu + dv}{au + bv}, \tag{7.120}$$

yielding $a = k_1$, $b = 0$, $c = 0$, and $d = k_2$. The radical, r, is computed as

$$r = (a + d)^2 + 4(bc - ad) = k_1^2 - 2k_1 k_2 + k_2^2 = (k_1 - k_2)^2. \tag{7.121}$$

Because $r > 0$, this singularity is either a NODE or a SADDLE. The singular point at $(0, 0)$ can be classified as a NODE since $(bc - ad) < 0$.

Singular point at $(k_2/k_4, k_1/k_3)$. The transformation is defined as

$$x = k_2/k_4 + u,$$
$$y = k_1/k_3 + v.$$

This produces the equation

$$\left.\frac{dv}{du}\right|_{(0,0)} = -\frac{k_2(y_s + v) - k_4(x_s + u)(y_s + v)}{k_1(x_s + u) - k_3(x_s + u)(y_s + v)}. \tag{7.122}$$

Substitution of $(x_s, y_s) = (k_2/k_4, k_1/k_4)$ and dropping the higher powers of u, v results in

$$\left.\frac{dv}{du}\right|_{(0,0)} = \frac{-k_1 k_4 u/k_3 + \text{HOT}}{-k_2 k_3 v/k_4 + \text{HOT}}, \tag{7.123}$$

which yields $a = 0$, $b = -k_1 k_3/k_4$, $c = -k_1 k_4/k_3$, and $d = 0$. The radical, r, is

$$r = (a + d)^2 + 4(bc - ad) = 4k_1 k_2. \tag{7.124}$$

Because $r > 0$, this singularity is either a NODE or a SADDLE. Since $(bc - ad) = 4k_1k_2 > 0$, the singular point at $(2, 0)$ can be classified as a SADDLE. It is left to the reader to establish the slope at this singularity as

$$s = \pm \frac{k_4}{k_3} \sqrt{\frac{k_1}{k_2}}. \tag{7.125}$$

EXAMPLE 7.8: PREDATOR-PREY INTERACTION (VOLTERRA'S MODEL)

This example is similar to the preceding one, but with a change of sign on one of the coefficients. This model was developed originally by Volterra to model population dynamics. The variable x represents the population of a species that lives on readily available food. The variable y represents the population of a species (predator) that eats x (prey). The birth and death rates of the prey species are modeled as proportional to its population, x. The prey's death rate also is assumed to be proportional to the population of the predator, y. The rate of change of the population of the prey is written analytically as

$$\frac{dx}{dt} = k_1 x - k_3 xy. \tag{7.126}$$

A similar model for the predator can be developed, but here the birth rate is assumed to be proportional to both the population of the predator and the prey whereas the death rate is only proportional to the predator's population. The new system of equations is now defined as

$$\frac{dy}{dt} = -k_2 y + k_4 xy. \tag{7.127}$$

In this model, all of the signs are explicitly given, and thus all of the coefficients are constrained to $k_i > 0$. It can be shown that the singular points are $(0, 0)$ and $(k_2/k_4, k_1/k_3)$, yielding singularities that are classified as a CENTER. The next question is whether the solution contours are closed. To simplify the analysis, normalizing transformations are introduced that force the singularity to $(1, 1)$:

$$w = \frac{k_4}{k_2} x, \tag{7.128}$$

$$z = \frac{k_3}{k_1} y. \tag{7.129}$$

The differential equations thus become

$$\dot{w} = k_1 w(1 - z), \tag{7.130}$$
$$\dot{z} = k_2 z(w - 1), \tag{7.131}$$

which, in canonical form, become

$$\frac{dw}{dz} = \frac{k_1 w(1 - z)}{k_2 z(w - 1)}. \tag{7.132}$$

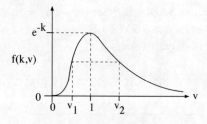

Figure 7.9. Functional $f(k, v) = e^{-kv}v^k$.

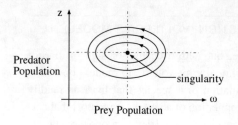

Figure 7.10. Relationship between predator and prey populations.

This relationship is separable and integrable. These operations yield

$$k_2 \ln w + k_1 \ln z = k_2 w + k_1 z + A, \tag{7.133}$$

which can be rewritten as

$$\{e^{-k_2 w} w^{k_2}\}\{e^{-k_1 z} z^{k_1}\} = e^A \doteq c^2. \tag{7.134}$$

This may be abstracted by introducing a functional, f, to obtain

$$f(k_2, w) f(k_1, z) = c^2, \qquad k_i > 0. \tag{7.135}$$

The functional has the interesting set of properties that $f(k, 0) = f(k, \infty) = 0$ and a maximum at $f(k, 1)$. This functional is shown in Fig. 7.9. From this figure, it can be observed that $c^2 \leq e^{-k_1} e^{-k_2}$, or $f(k_2, w) f(k_1, z) \leq e^{-(k_1+k_2)}$, which infers that the solution contours are all closed, as shown in Fig. 7.10. This system therefore is classified as a conservative nonlinear system. It is left as an exercise for the reader to determine that the singularity at the origin is a SADDLE and the singularity at (1, 1) is a CENTER.

7.3 POINCARÉ INDEX

After the success of Example 7.8, it would be natural to ask if it were possible to predict the presence of limit cycles from a singular point analysis. This generally is not possible, but it frequently may be shown that a limit cycle does not exist. The definition of the Poincaré index is defined as

> For a vector field defined in a two-dimensional region, R, and containing a simple closed path, N, the Poincaré index of the region enclosed by N is the number of revolutions of the vector along N for one complete circuit.

If N encloses a singular point, (x_s, y_s), then this region inherits the index value of this singularity, provided no other singularities are contained within N. Consider the singular

7.3. Poincaré Index

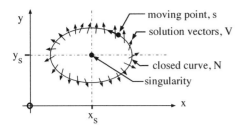

Figure 7.11. Closed path around a singularity.

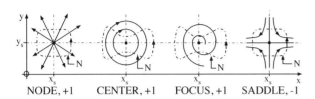

Figure 7.12. Poincaré index of singularities.

point and solution trajectories shown in Fig. 7.11. The method of computing the Poincaré index is given as

1. Draw a closed path, N, around the singularity.
2. Let a point S on N move counterclockwise around N.
3. Let a vector V that points in the direction of the solution and that lies on S be carried along as S moves on N.
4. The number of revolutions of the vector V for a complete circuit of N is the Poincaré index.

This technique may be applied to the four types of singularities as depicted in Fig. 7.12. The corresponding Poincaré indices of NODE, SADDLE, CENTER, and FOCUS singularities are $+1$, -1, $+1$, and $+1$, respectively. This method thus discriminates between SADDLE (index $=-1$) and non-SADDLE (index $=+1$) singular points. It also has been observed that the Poincaré index is zero if there are no singularities within the contour, N. If there are multiple singularities, the resulting Poincaré index is the sum of the individual indices.

An analytical method for computing the Poincaré index, $\mathcal{PI}$, may be derived. As in Eqs. (7.23) and (7.24), the terms P and Q are defined as

$$\left. \begin{array}{l} \frac{dx}{dt} = P(x,y) \\ \frac{dy}{dt} = Q(x,y) \end{array} \right\}. \tag{7.136}$$

Thus, their ratio is

$$\frac{dy}{dx} = \frac{Q(x,y)}{P(x,y)} = \tan\theta, \tag{7.137}$$

where θ is the direction of the vector V. The Poincaré index then is found by integrating θ around N, or

$$\mathcal{PI} = \frac{1}{2\pi} \oint_N d\theta = \frac{1}{2\pi} \oint_N d\left[\tan^{-1}\left(\frac{Q(x,y)}{P(x,y)}\right)\right]. \tag{7.138}$$

This may be reduced to

$$\mathcal{PI} = \frac{1}{2\pi} \oint_N \left[\frac{P\,dQ - Q\,dP}{P^2 + Q^2} \right]. \tag{7.139}$$

Consider a linear system as described in Eq. (7.136), having a singularity at (0, 0) and two real characteristic roots (λ_1, λ_2) or a pair of complex roots $u \pm jv$, where $u \geq 0$, $v > 0$, and $j = \sqrt{-1}$. It is possible to find a linear, reversible transformation to cast the system into normal form as in systems e_1 and e_2:

$$e_1: \begin{array}{l} \frac{dx}{dt} = \lambda_1 x \\ \frac{dy}{dt} = \lambda_2 y \end{array} \quad \text{or} \quad e_2: \begin{array}{l} \frac{dx}{dt} = ux - vy \\ \frac{dy}{dt} = vx + uy \end{array}. \tag{7.140}$$

For a closed path, N, defined by the four lines $x = \pm 1$ and $y = \pm 1$, for system e_1,

$$\begin{aligned} \mathcal{PI} &= \frac{1}{2\pi} \oint_N \frac{\lambda_1 \lambda_2 (x\,dy - y\,dx)}{\lambda_1^2 x^2 + \lambda_2^2 y^2} \\ &= \frac{1}{2\pi} \left[\oint_{-1}^{1} \frac{\lambda_1 \lambda_2\,dy}{\lambda_1^2 + \lambda_2^2 y^2} - \oint_{1}^{-1} \frac{\lambda_1 \lambda_2\,dy}{\lambda_1^2 + \lambda_2^2 y^2} - \oint_{-1}^{1} \frac{\lambda_1 \lambda_2\,dx}{\lambda_1^2 x^2 + \lambda_2^2} + \oint_{1}^{-1} \frac{\lambda_1 \lambda_2\,dx}{\lambda_1^2 x^2 + \lambda_2^2} \right] \\ &= \begin{cases} +1 & \lambda_1 \lambda_2 \geq 0 \\ -1 & \lambda_1 \lambda_2 < 0 \end{cases}. \end{aligned}$$

Similarly, for system e_2

$$\begin{aligned} \mathcal{PI} &= \frac{1}{2\pi} \oint_N \frac{(ux - vy)(v\,dx + u\,dy) - (vx + uy)(u\,dx - v\,dy)}{(ux - vy)^2 + (vx + uy)^2} \\ &= \frac{1}{2\pi} \oint_N \frac{x\,dy - y\,dx}{x^2 + y^2} = 1. \end{aligned}$$

This analytical method therefore produces results identical to those obtained using the pictorial method.

The Poincaré index for any region that does not contain any singularities is considered next. If any functions F and G are continuous within N and their partial derivatives with respect to x and y exist within N, then Green's theorem states that

$$\iint_D \left(\frac{\partial G}{\partial x} - \frac{\partial F}{\partial y} \right) dx\,dy = \oint_N (F\,dx + G\,dy). \tag{7.141}$$

Rewriting Eq. (7.139) to appear more like Green's formulation,

$$\begin{aligned} \mathcal{PI} &= \frac{1}{2\pi} \oint_N \frac{P\,dQ - Q\,dP}{P^2 + Q^2} \\ &= \frac{1}{2\pi} \left[\oint_N \frac{P\partial Q/\partial x - Q\partial P/\partial x}{P^2 + Q^2} dx + \frac{P\partial Q/\partial y - Q\partial P/\partial y}{P^2 + Q^2} dy \right]. \end{aligned} \tag{7.142}$$

It is assumed that the first and second partial derivatives of P and Q exist within N. If we define

$$F(x, y) = \frac{P\partial Q/\partial x - Q\partial P/\partial x}{P^2 + Q^2}, \qquad G(x, y) = \frac{P\partial Q/\partial y - Q\partial P/\partial y}{P^2 + Q^2}, \tag{7.143}$$

7.3. Poincaré Index

then Green's theorem may be applied to Eq. (7.142) to obtain

$$\mathcal{PI} = \oiint_D \left(\frac{\partial G}{\partial x} - \frac{\partial F}{\partial y} \right) dx\, dy = 0. \tag{7.144}$$

From these properties, the following corollaries may be obtained:

1. The Poincaré index of a singular point is independent of N. Thus, any two paths containing the same singular point will have the same Poincaré index values.
2. For any path, N, containing k singular points, the Poincaré index of N is the sum of the indices of the k points.
3. An essential condition for a limit cycle to exist at a singular point is that the Poincaré index be $+1$.

EXAMPLE 7.9: CUBIC SPRING-MASS STABILITY

Consider the differential equation

$$\ddot{x} + x^3 = 0, \qquad x(0) = x_0. \tag{7.145}$$

If we define $v = \dot{x}$, this produces the equation set

$$\dot{x} = v,$$
$$\dot{v} = -x^3,$$

or

$$\frac{dv}{dx} = -\frac{x^3}{v} = \frac{cx + dv}{ax + bv}. \tag{7.146}$$

This equation has a singular point at $(0, 0)$; linearizing the equation around the origin and solving for the coefficients yields $a = 0$, $b = -1$, $c = 0$, and $d = 0$. However, the restriction that $bc \neq ad$ means that standard singular point analysis cannot be applied. *Simple singular points* are those where the function may be linearized.

Fortunately, in this example an analytic solution for $v(x)$ may be obtained by integrating Eq. (7.146) directly to obtain

$$v = \pm\sqrt{v_0^2 - (x^4 - x_0^4)/8}. \tag{7.147}$$

Plotting this function yields closed limit cycles. This is a *nonsimple singular point*, which infers that the Poincaré index is not necessarily ± 1, but in this instance, $\mathcal{PI} = +1$.

In summary, three theorems are presented regarding the Poincaré index.

Theorem 1. Green's theorem for plane curves can be used to show that, provided P, Q, P', and Q' are continuous and not simultaneously zero (no singular points) within the curve N, $\mathcal{PI} = 0$.

Theorem 2. If the smoothness of P, Q is satisfied in any region containing one singularity, then any simple closed curve, N, generates the same $\mathcal{PI}$.

Theorem 3. If p is the number of times that Q/P changes from $+\infty$ to $-\infty$, and n is the number of times it changes from $-\infty$ to $+\infty$ on the curve N, then $\mathcal{PI} = (p-n)/2$.

7.4 BENDIXSON'S FIRST THEOREM

This technique was introduced to define situations in which limit cycles do not exist. Bendixson stated that there is no limit cycle in any given region if any of the following conditions hold [11]:

1. There are no singular points in the region.
2. There is only one singular point and it has a Poincaré index of -1.
3. There are several singular points, but no collection of them has a Poincaré index of $+1$.

EXAMPLE 7.10: POINCARÉ INDEX OF A VAN DER POL OSCILLATOR

Consider the differential equation for a van der Pol oscillator:

$$\ddot{x} + \mu(x^2 - 1)\dot{x} + x = 0. \tag{7.148}$$

For $\mu > 0$, there is either a NODE or a FOCUS at the origin and the index is $\mathcal{PI} = +1$. One can try many different closed regions, but the application of Bendixson's theorem leads to the conclusion that there are no limit cycles.

An alternative version of Bendixson's theorem is stated as

No limit cycle exists in any region in which the term $(\partial P/\partial x + \partial Q/\partial y)$ does not change sign.

This may be clarified through the following example.

EXAMPLE 7.11: ALTERNATIVE BENDIXSON THEOREM TEST OF A VAN DER POL OSCILLATOR

Consider the formulation of the van der Pol oscillator given as

$$\dot{x} = y = P(x, y),$$
$$\dot{y} = -x - \mu(x^2 - 1)y = Q(x, y),$$

which results in

$$\left(\frac{\partial P}{\partial x} + \frac{\partial Q}{\partial y}\right) = \mu(1 - x^2). \tag{7.149}$$

This term does not change sign within the range $0 \leq |x| \leq 1$ and in this region, no limit cycle may exist. Interestingly, limit cycles do exist for van der Pol oscillators, as depicted in Fig. 7.13, but they can only exist outside of this region.

7.4. Bendixson's First Theorem

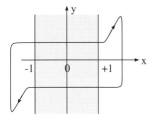

Figure 7.13. Phase-plane solution for a van der Pol oscillator.

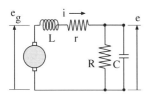 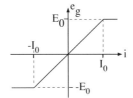

Figure 7.14. Schematic of generator and its operational behavior.

It is possible to have singular points for n-dimensional problems. In this case, the differential equations assume the form

$$\dot{x}_1 = P(x_i),$$
$$\dot{x}_2 = Q(x_i),$$
$$\vdots = \vdots$$
$$\dot{x}_n = R(x_n).$$

For a singularity to exist, $P, Q, \ldots, R$ must all vanish simultaneously.

EXAMPLE 7.12: D.C. GENERATOR STABILITY

Consider the system depicted in Fig. 7.14. The governing equations for this system are

$$e = e_g - L\frac{di}{dt} - ri, \tag{7.150}$$

$$i = \frac{e}{R} + C\frac{de}{dt}, \tag{7.151}$$

where e = load voltage, e_g = generated voltage, i = load current, r = internal generator resistance, L = self-inductance of generator, R = load resistance, and C = capacitance of load.

The differential equations may be reduced (normalized) by introducing the scaling equations

$$\left.\begin{array}{l} x = \frac{i}{I_0} \\ y = \frac{e}{E_0} \\ \tau = \frac{rt}{L} \\ R_0 = \frac{E_0}{I_0} \end{array}\right\}. \tag{7.152}$$

To further simplify the algebra, the following positive-definite symbols are introduced:

$$\left.\begin{array}{l} \alpha = r/R_0 \\ \beta = R/R_0 \\ \gamma = rRC/L \end{array}\right\}. \tag{7.153}$$

The generator's performance function, $f(x)$, also is normalized to ± 1:

$$f(x) = \frac{e_g I_0 r}{E_0} = \begin{cases} x & |x| \le 1 \\ 1 & x > 1 \\ -1 & x < 1 \end{cases}. \tag{7.154}$$

When Eqs. (7.152), (7.153), and (7.154) are substituted into the governing equations, this produces the scaled transformed equations for the system

$$\frac{dx}{d\tau} = \frac{1}{\alpha}[f(x) - y] - x, \tag{7.155}$$

$$\frac{dy}{d\tau} = \frac{1}{\gamma}[\beta x - y], \tag{7.156}$$

which thereby produce the phase slope function,

$$\frac{dy}{dx} = \frac{\frac{1}{\gamma}[\beta x - y]}{\frac{1}{\alpha}[f(x) - y]}$$

$$= \frac{\frac{\beta x}{\gamma} - \frac{1}{\gamma} y}{x(\frac{1}{\alpha} - 1) - \frac{1}{\alpha} y}$$

$$= \frac{cx + dy}{ac + by}.$$

Singular point near origin. The singular point at the origin is in the region where $f(x) = x$. The phase-plane slope is then

$$\frac{dy}{dx} = \frac{\frac{\beta x}{\gamma} - \frac{y}{\gamma}}{x(\frac{1}{\alpha} - 1) - \frac{y}{\alpha}} \tag{7.157}$$

$$= \frac{cx + dy}{ac + by}. \tag{7.158}$$

From this equation, the stability parameters are $a = (1-\alpha)/\alpha$, $b = -1/\alpha$, $c = b/\gamma$, and $d = -1/\gamma$. Attempting to compute $(a+d)^2 + 4(bc-ad)$ is an algebraic tangle. Another procedure is to return to the characteristic equation given as

$$\begin{vmatrix} (a-\lambda) & b \\ c & (d-\lambda) \end{vmatrix} = 0, \tag{7.159}$$

from which the characteristic roots may be found from the equation

$$\lambda^2 + \left[1 - \frac{1}{\alpha} + \frac{1}{\gamma}\right]\lambda + \left[\frac{\beta}{\alpha\gamma} + \frac{1}{\gamma}\left(1 - \frac{1}{\alpha}\right)\right] = 0. \tag{7.160}$$

To ensure stability, the characteristic roots cannot change sign. Therefore, the following terms both must be greater than zero:

$$1 - \frac{1}{\alpha} + \frac{1}{\gamma} > 0, \tag{7.161}$$

$$\frac{\beta}{\alpha\gamma} + \frac{1}{\gamma}\left(1 - \frac{1}{\alpha}\right) > 0. \tag{7.162}$$

7.4. Bendixson's First Theorem

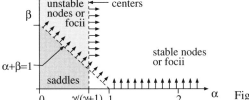

Figure 7.15. Stability graph for the generator.

These relations therefore can be used as constraints on the stability of the system

$$\alpha = \frac{\gamma}{\gamma+1}, \tag{7.163}$$

$$\alpha + \beta = 1 \tag{7.164}$$

and depicted schematically as in Fig. 7.15. Equation (7.160) can be written as

$$\lambda^2 + A\lambda + B = 0, \tag{7.165}$$

yielding characteristic roots

$$\lambda = \tfrac{1}{2}\left[-A \pm \sqrt{A^2 - 4B}\right]. \tag{7.166}$$

The resulting singular points then could be classified according to the values of A, B. This is shown as

Parameters	Type
$A = 0$, $B > 0$	CENTER
$A = 0$, $B < 0$	SADDLE
$A > 0$, $B > 0$	NODE or FOCUS (stable)
$A > 0$, $B < 0$	SADDLE
$A < 0$, $B > 0$	NODE or FOCUS (unstable)
$A < 0$, $B < 0$	SADDLE

Singular points not at origin. For these singularities, the generator function takes on the values $f(x) = \pm 1$ and the slope on the phase plane is

$$\frac{dy}{dx} = \frac{\frac{1}{\gamma}[\beta x - y]}{\frac{1}{\alpha}[\pm 1 - y] - x}. \tag{7.167}$$

The resulting singular points are thus only along the lines $y = \beta x$, and $x = [\pm 1 - y]/\alpha$. These reduce to the values

$$y = \frac{\pm \beta}{\alpha + \beta}, \tag{7.168}$$

$$x = \frac{\pm 1}{\alpha + \beta}, \tag{7.169}$$

TABLE 7.2. Poincaré Index, $\mathcal{PI}$, for the D.C. Generator Example

Singularity location	$\mathcal{PI}$	Singularity type
$(0, 0)$	-1	SADDLE
$(+x_s, +y_s)$	$+1$	NODE or FOCUS
$(-x_s, -y_s)$	$+1$	NODE or FOCUS
$\sum \mathcal{PI} =$	$+1$	Total Index Change

but only for the region $|x| > 0$. Therefore, $\alpha + \beta < 1$. Translation to the "+" singular point yields

$$\frac{dv}{du} = \frac{\frac{\beta u}{\gamma} - \frac{v}{r}}{-u - \frac{v}{\alpha}} = \frac{cu + dv}{au + bv}. \tag{7.170}$$

The characteristic roots therefore are found from the characteristic equation,

$$\lambda^2 + \left(1 + \frac{1}{\gamma}\right)\lambda + \left(\frac{\beta}{\alpha\gamma} + \frac{1}{\gamma}\right) = 0. \tag{7.171}$$

The terms α, β, and γ are all positive definite, and the characteristics are stable since there are no sign changes. The singular points are of type NODE. The damping rules out CENTERs and SADDLEs, which are unstable.

A limit-cycle investigation of this example can be undertaken using the Bendixson method. The governing differential equations for $|x| < 1$ are

$$\frac{dx}{dt} = P(x, y) = \frac{1 - \alpha}{\alpha}x - \frac{1}{\alpha}y, \tag{7.172}$$

$$\frac{dy}{dt} = Q(x, y) = \frac{\beta}{\gamma}x - \frac{1}{\gamma}y, \tag{7.173}$$

and we must examine the terms

$$\frac{\partial P}{\partial x} + \frac{\partial Q}{\partial y} = \frac{1}{\alpha} - 1 - \frac{1}{\gamma}, \qquad |x| < 1, \tag{7.174}$$

$$\frac{\partial P}{\partial x} + \frac{\partial Q}{\partial y} = -1 - \frac{1}{\gamma}, \qquad |x| > 1. \tag{7.175}$$

By examining Eqs. (7.174) and (7.175), it is found that, for $|x| < 1$ and $1/\alpha - 1 - 1/\gamma < 0$, there will never be a sign change and no limit cycles can exist. This is depicted in Fig. 7.15. Consider the calculation of the Poincaré index for this example, as shown in Table 7.2. Since the sum of the indices is $+1$, there exists the possibility of a limit cycle enclosing all three singularities.

7.5 SECOND METHOD OF LYAPUNOV

This work was originally introduced in Russia [28] by Lyapunov,[2] a student of Tchebycheff for whom the Tchebycheff polynomials were named. In this method, a system is described

---

[2] A.M. Lyapunov (1857–1918).

7.5. Second Method of Lyapunov

by the set of equations

$$\dot{x}_1 = X_1(x_1, x_2, x_3, \ldots, x_n),$$
$$\dot{x}_2 = X_2(x_1, x_2, x_3, \ldots, x_n),$$
$$\vdots \quad \vdots$$
$$\dot{x}_n = X_n(x_1, x_2, x_3, \ldots, x_n),$$

where x_i are the state variables and the X_i are functions of these variables. This analysis is restricted to those functions where

$$X_j(0, 0, \ldots, 0) = 0. \tag{7.176}$$

Using the method of Lyapunov, there are four possible results:

1. The system is asymptotically stable.
2. The system is stable in the Lyapunov sense.
3. The system is unstable.
4. The result is inconclusive.

In the sections that follow, a Lyapunov function is defined, followed by a description of Lyapunov's first and second theorems.

7.5.1 Lyapunov Function

A new functional, V, is introduced here which has the following properties:

1. $V(x_i = 0) = 0$ and V is continuous around the origin.
2. V is termed negative/positive definite only if it is zero at the origin and negative/positive everywhere else.
3. V is termed negative/positive semidefinite if it has the same sign $(+/-)$ throughout the region except that it is zero at a number of points, including the origin.
4. V is termed indefinite if it is not sign-definite and yields inconclusive answers.

Some examples of the Lyapunov function are

$$V(x_1, x_2) = x_1^2 + x_2^2 \rightarrow \text{Positive Definite},$$
$$V(x_1, x_2, x_3) = x_1^2 + x_2^2 \rightarrow \text{Positive Semidefinite},$$
$$V(x_1, x_2) = x_1 + x_2 \rightarrow \text{Indefinite}.$$

These functions are to be used to determine the stability of systems. There is no specific rational basis for the choice of the Lyapunov functions, but it is shown that one generally chooses a negative-definite function for stability analysis. The proper choice of this function is based on its use in Lyapunov's stability theorem given in the following section. Often, one chooses this function on the basis of mechanical energy considerations (e.g., $V =$ potential energy + kinetic energy).

7.5.2 First Theorem of Lyapunov

This theorem is the most general nonlinear system stability criterion available. It is the cornerstone of modern stability theory and is stated as

> If $V(x_i)$ is continuous through all the first partial derivatives, and $V(x_i = 0) = 0$, and $V(x_i) > 0$ for all $|x_i| > 0$, then the system is stable (may contain a limit cycle) in the region containing the origin, where $dv/dt \leq 0$, $|x_i| > 0$, or the system is asymptotically stable in the region containing the origin where $dv/dt < 0$, $|x_i| > 0$.

EXAMPLE 7.13: LINEAR SPRING-MASS SYSTEM

The simple spring-mass system may be represented as

$$\dot{x}_1 = x_2,$$
$$\dot{x}_2 = -\frac{k}{m}x_1.$$

The Lyapunov V function chosen is $V = x_1^2 + x_2^2$, which has derivative

$$\frac{dV}{dt} = \frac{\partial V}{\partial x_1}\frac{dx_1}{dt} + \frac{\partial V}{\partial x_2}\frac{dx_2}{dt} = 2x_1\dot{x}_1 + 2x_2\dot{x}_2. \quad (7.177)$$

Inserting the original differential equations yields

$$\frac{dV}{dt} = 2\left[x_1 x_2 - x_1 x_2 \frac{k}{m}\right] = 2x_1 x_2 \left[1 - \frac{k}{m}\right]. \quad (7.178)$$

An analysis of this functional does not produce any conclusive result regarding the stability of the system. This means that information cannot be derived from this functional and another must be chosen. A second V function is selected as $V = x_1^2 + Ax_2^2$. The derivative becomes

$$\frac{dV}{dt} = 2x_1 x_2 \left[1 - A\frac{k}{m}\right]. \quad (7.179)$$

By choosing the values of the arbitrary constant $A = m/k$, then $dV/dt = 0$ everywhere and the system is Lyapunov stable.

EXAMPLE 7.14: NONLINEAR SPRING-MASS SYSTEM

Consider the system described by

$$M\frac{d^2y}{dt^2} + Ky + Dy^3 = 0, \quad (7.180)$$

where $M, K, D > 0$. A singular point analysis of this system portrays the singularity at the origin as a CENTER. Lyapunov's method also can be used to analyze the stability of

7.5. Second Method of Lyapunov

the system. For this system, the functional chosen is $V = x_1^2 + Ax_2^2$, subject to $A > 0$, as in Example 7.13. The system may be written as

$$\dot{x}_1 = x_2,$$
$$\dot{x}_2 = -\frac{K}{M}x_1 - \frac{D}{M}x_1^3.$$

Through selection of $A = m/k$, the derivative of the functional becomes

$$\frac{dV}{dt} = -2\frac{D}{K}x_1^3 x_2, \qquad (7.181)$$

which is sign indefinite and therefore inconclusive. Another functional is selected as

$$\frac{KV}{2} = \frac{Kx_1^2}{2} + \frac{mx_2^2}{2}, \qquad (7.182)$$

or another form of energy function,

$$V = \int_0^{x_1} (Ky + Dy^3) dy + \frac{Mx_2^2}{2} = \frac{Kx_1^2}{2} + \frac{Dx_1^4}{4} + \frac{Mx_2^2}{2}, \qquad (7.183)$$

which yields $dV/dt = 0$. Thus, here is a V functional that is positive definite and is a conservative (globally consistent) system.

EXAMPLE 7.15: LINEAR SPRING-MASS-DAMPER SYSTEM

Consider the system described by

$$M\frac{d^2y}{dt^2} + C\frac{dy}{dt} + Ky = 0, \qquad (7.184)$$

where $M, K, C > 0$. A singular point analysis of this system portrays the singularity at the origin as a CENTER. For simplicity, the constants A, B are defined as $A = C/M$ and $B = K/M$, which yields

$$\frac{d^2y}{dt^2} + A\frac{dy}{dt} + By = 0. \qquad (7.185)$$

The V functional chosen is positive definite,

$$V = a_{11}x_1^2 + a_{12}x_1x_2 + a_{22}x_2^2, \qquad (7.186)$$

provided that the a_{ij} values are positive, $a_{ij} > 0$, and if

$$a_{11}a_{22} - a_{12}^2 > 0 \qquad (7.187)$$

(refer to Perli [32], p. 95). The system of first-order equations produced is thus

$$\dot{x}_1 = x_2,$$
$$\dot{x}_2 = -Bx_1 - Ax_2,$$

and results in the functional derivative

$$\frac{dV}{dt} = (2a_{11}x_1 + 2a_{12}x_2)\dot{x}_1 + (2a_{12}x_1 + 2a_{22}x_2)\dot{x}_2$$
$$= x_1^2(-2a_{12}B) + x_2^2(-2a_{22}A + 2a_{12}) + x_1x_2(2a_{11} - 2a_{12}A - 2a_{22}B).$$

If the attempt is made to constrain the system to be negative definite of the form

$$\frac{dV}{dt} = -(x_1^2 + x_2^2), \tag{7.188}$$

then this implies that

$$-2a_{12}B = -1,$$
$$-2a_{22}A + 2a_{12} = -1,$$
$$2a_{11} - 2a_{12}A - 2a_{22}B = 0,$$

which can be solved to obtain

$$a_{12} = \frac{1}{2B} = \frac{M}{2K},$$
$$a_{22} = \frac{B+1}{2AB} = \frac{M(K+M)}{2KC},$$
$$a_{11} = \frac{A}{2B} + \frac{B+1}{2A} = \frac{C}{2M}\left(\frac{K}{M} + \left(\frac{K}{M}+1\right)\right).$$

The constraint of Eq. (7.187) is thus

$$a_{11}a_{22} - a_{12}^2 = \frac{1}{4B} + \frac{(B+1)^2}{4A^2B} > 0. \tag{7.189}$$

Thus, the system is globally asymptotically stable.

EXAMPLE 7.16: LINEAR SPRING-MASS-DASHPOT REVISITED

Consider the linear system described by the governing differential equation

$$m\ddot{y} + c\dot{y} + ky = 0, \quad m, k, c > 0. \tag{7.190}$$

As in Example 7.15, the origin is recognized as a stable FOCUS. If the problem is redefined as two first-order systems by assuming $x_1 = y$, $\dot{x}_1 = x_2$, and the V function chosen is

$$V = \frac{kx_1^2}{2} + \frac{mx_2^2}{2}, \tag{7.191}$$

then the first derivative of the V function becomes

$$\frac{dV}{dt} = kx_1\dot{x}_1 + mx_2\dot{x}_2, \tag{7.192}$$

7.5. Second Method of Lyapunov

which reduces to

$$\frac{dV}{dt} = -cx_2^2. \qquad (7.193)$$

This yields the result that the system is Lyapunov stable. It is known that the system is globally asymptotically stable; therefore, all information cannot be derived from a single V function. Barbashin and Krasovski [4] state that a system is asymptotically stable in some region about the origin if, within this region, a positive-definite V function may be found such that

- dV/dt is negative semidefinite or definite in this region, and
- contours where $dV/dt = 0$ in this region are not solutions of the system equation.

For the current example, $dV/dt = -cx_2^2$, which is negative semidefinite everywhere. The only locations where this derivative vanishes are along the line $x_2 = 0$, but these are not solution trajectories. Thus, the system in question is asymptotically stable.

7.5.3 Second Theorem of Lyapunov

This theorem states that if there exists a function

$$\frac{dV}{dt} = W(x_1, x_2, \ldots, x_n), \qquad (7.194)$$

such that

1. W is continuous,
2. at the origin, $W(0, 0, \ldots, 0) = 0$, and
3. $W(x_i) < 0$ for all $|x_i| > 0$,

then the system is *unstable* in the region containing the origin of the phase space for which the inequality $V(x_1, x_2, \ldots, x_n) > 0$ is not satisfied.

EXAMPLE 7.17: SIMPLE CONTROLLER

Consider the system shown in Fig. 7.16. For this system, $\epsilon = x - y$ and

$$\int (k\epsilon)\, dt = y \qquad (7.195)$$

or

$$k\epsilon = \frac{dy}{dt} = \dot{y}. \qquad (7.196)$$

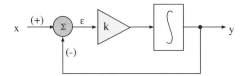

Figure 7.16. Simple linear controller.

If x is assumed to be constant, then

$$\dot{\epsilon} = -\dot{y} \tag{7.197}$$

or

$$\dot{\epsilon} = -k\epsilon. \tag{7.198}$$

For this system, the V function chosen will be of the form $V = -x_1^2$, where $x_1 = \epsilon$. The first derivative then becomes

$$\frac{dV}{dt} = -2x_1\dot{x}_1 = W(x_1), \tag{7.199}$$

but $\dot{x}_1 = -kx_1$, thus,

$$W = \frac{dV}{dt} = +2kx_1^2. \tag{7.200}$$

Consider the two possible cases for W:

$k < 0$. Since V is never positive and $W < 0$ for all $|x_i| > 0$ with W continuous and $W(0) = 0$, the requirements of the instability theorem are met. Since $V > 0$ is not satisfied for any x_1, the system is totally unstable.

$k > 0$. With W now positive definite, V is negative definite. Since these two entities have opposite signs, the system is asymptotically stable.

EXAMPLE 7.18: NONLINEAR CONTROLLER

Consider the nonlinear system shown in Fig. 7.17. The function $f(\epsilon)$ has the properties

$$f(\epsilon) = 0, \quad \epsilon = 0,$$
$$\epsilon f(\epsilon) > 0, \quad |\epsilon| > 0,$$

resulting in $f(\epsilon)$ always being in the first and third quadrant. Possible V functions are

$$V = \int_0^{x_1} f(t)\, dt,$$
$$V = x_1^2 \quad \text{where} \quad x_1 = \epsilon.$$

Choosing the latter function yields

$$W = \frac{dV}{dt} = 2x_1\dot{x}_1 = -2x_1 f(x_1). \tag{7.201}$$

This system thus is asymptotically stable for all constant inputs and initial conditions.

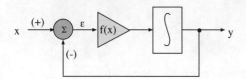

Figure 7.17. Nonlinear controller.

7.6 INTRODUCTION TO CHAOTIC SYSTEMS

The stability of nonlinear systems may be a function of the operating conditions for the system, not factors obvious in the system itself. Poincaré, perhaps the greatest French mathematician in the second half of the nineteenth century, noted that minor variations in the initial conditions of some nonlinear systems resulted in large differences in their long-term solutions. This irregular response of a class of nonlinear systems is termed *chaotic*. Although chaotic response may appear to result from stochastic effects, the phenomena are fundamentally different. Stochastic systems are those that have random external forces acting on them and that result in irregular behavior. Chaotic response is a character inherent in the dynamics of the system.

Baker and Gollub [3] state that there are only two essential criteria that must be met for a system to exhibit chaotic behavior:

1. The governing equations must be nonlinear and contain nonlinear terms coupling several variables.
2. The system must have at least three independent dynamic variables. Although this infers that the order of the systems involved must be three or greater. The existence of an independent forcing function may satisfy this requirement.

The differential equations satisfying these criteria are of the form

$$\frac{dx_1}{dt} = F_1(x_1, x_2, \ldots, x_n),$$
$$\frac{dx_2}{dt} = F_2(x_1, x_2, \ldots, x_n),$$
$$\vdots \quad \vdots$$
$$\frac{dx_n}{dt} = F_n(x_1, x_2, \ldots, x_n).$$

where n must be greater than or equal to 3 and the forcing functions, F_i, are nonlinear functions of the dependent variables, x_i.

One of the most common nonlinear systems described in the literature is the simple pendulum. Although the pendulum is thought to be the epitome of regularity, when one adds a harmonic forcing function, the system may exhibit chaotic behavior. The governing equations for the nonforced pendulum are stated in Section 2.5.3.

Consider the governing differential equation for a damped pendulum with an external forcing function:

$$mL^2 \frac{d^2\theta}{dt^2} + \gamma \frac{d\theta}{dt} + WL \sin(\theta) = A \cos(\omega_D t). \tag{7.202}$$

This may be rewritten as

$$\frac{d\omega}{dt} = -\frac{1}{q}\omega - \sin\theta + g\cos\phi, \tag{7.203}$$

$$\frac{d\theta}{dt} = \omega, \tag{7.204}$$

$$\frac{d\phi}{dt} = \omega_D. \tag{7.205}$$

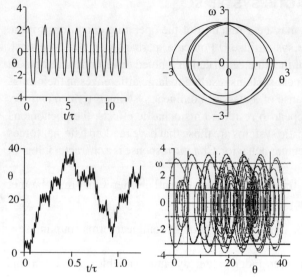

Figure 7.18. Transient response and limit cycle of a damped, forced pendulum: (a) Pendulum angle vs time, $q = 2$, $g = 1.5$, $\theta_0 = 2$, $\omega_0 = 0$; (b) Phase plot, pendulum velocity vs angle.

Figure 7.19. Chaotic response of a damped, forced pendulum: (a) Pendulum angle vs time, $q = 2$, $g = 3.1$; (b) Phase plot, pendulum velocity vs angle.

The introduction of the phase variable, $\phi(t)$, along with the velocity, ω, converts this second-order problem into three first-order equations which satisfy the criteria defined for chaotic systems. This does not assure that the system is chaotic, but these are necessary conditions. Thus, these criteria are necessary but not sufficient conditions for chaotic behavior.

For certain values of the constants in Eqs. (7.203) and (7.205), the pendulum exhibits chaotic behavior. Figure 7.18 depicts the nonchaotic response of a damped, forced pendulum; Fig. 7.19 portrays the same system with a different forcing function amplitude, g. Note that the chaotic response of the pendulum has no apparent limit cycle, but it appears to have two dominant frequencies.

7.6.1 Phase Space

In Chapter 5, Section 5.1, the phase plane for a pendulum was introduced as a two-dimensional plot of angular velocity versus position. As introduced in Eqs. (7.203) to (7.205), however, the true phase space for a forced pendulum is the three-space ω, θ, ϕ. The third dimension is constantly increasing with time at a rate dependent on the frequency, ω_D, of the forcing function.

In the phase plane, there are several properties that real systems exhibit and that provide clues to the stability of the system.

No-crossing property. Adjacent trajectories in phase space may never cross. These trajectories may approach one another asymptotically, but they can never cross. A crossing of any two trajectories would introduce uncertainty or ambiguity into the solution of the systems. Systems that have this property of crossing trajectories are *indeterminate* and are not of general interest to the modeler. Mechanical systems are subject to the laws of statics, dynamics, and thermodynamics and their next state is always known, given the current state. They are thus *determinate* systems.

7.6. Introduction to Chaotic Systems

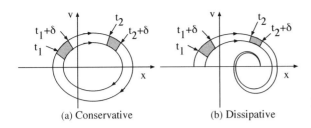

Figure 7.20. Trajectory cross-sectional areas.

(a) Conservative   (b) Dissipative

Cross-sectional properties. It is possible to classify systems on the basis of their phase-plane behavior as either conservative or dissipative. In conservative systems, such as constant-energy dynamic mechanical systems, the areas between two trajectories will remain constant. In most such systems, the solutions are limit cycles, as shown in Fig. 7.20(a). In this figure, two separate areas at different times for two trajectories also are shown to portray the constant-area concept. The group of points corresponding to area 1 would later be identified as area 2. For conservative systems, these two areas remain the same.

In dissipative systems, the cross-sectional area will diminish with time. This is depicted in Fig. 7.20(b). The solutions for long times may approach a stable solution known as an attractor point. The center of a limit cycle is also an attractor point. Such points are also termed singularities (see Section 7.2).

7.6.2 Poincaré Sections

A valuable tool in understanding the behavior of chaotic systems is a modification of the phase-plane graph. As demonstrated by the driven pendulum behavior in Fig. 7.19, chaotic behavior produces phase-plane plots that appear much like the scribbling of a young child. What must be remembered, however, is that the systems being discussed in this text are determinate and they behave according to natural laws. In spite of the apparent chaos, if one could measure the position, velocity, and the forcing function's amplitude and phase, it would be possible to predict its subsequent motion for some period of time. If our analysis procedures were exact, we could predict the subsequent motion for any future time precisely. Otherwise, the predictions are limited by the precision of the computing system used and any approximations inherent in the analysis procedure.

Poincaré invented a technique to bring out hidden structure within the solution by periodically sampling the result to produce a form of analytical stroboscope. If one uses the period of the forcing function to sample the result and plots this on the phase plane, a single point appears. This plot is termed a Poincaré section. If the parameters of the system or the frequency of the forcing function are changed, the resulting Poincaré section portrays some interesting behavioral patterns for the system. A typical Poincaré section is shown as Fig. 7.21. As an historical note, one should greatly appreciate the insight and innovation of Poincaré, who invented the section concept. Each of the single points shown in Fig. 7.21 required several hundred numerical computations. To have envisaged that this structure even existed, given the tools available at the turn of the century, is a remarkable feat.

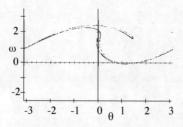

Figure 7.21. Poincaré section for a damped, forced pendulum.

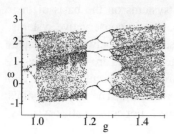

Figure 7.22. Bifurcation diagram for the sensitivity of a damped pendulum to driving amplitude, g.

7.6.3 Bifurcation Diagrams

Another valuable analytical tool in understanding chaotic systems is the bifurcation diagram. A bifurcation is when the number of solutions to the governing differential equation for a system changes as a parameter changes. Consider the nonlinear, damped, and driven pendulum described at the beginning of this chapter as the parameter q in Eq. (7.203) is varied. If the solution is allowed to continue long enough to achieve a limit cycle for each value of q tested, then a plot of the limit-cycle frequency, ω, at a fixed phase angle of the solution versus q is as shown in Fig. 7.22. What is interesting here is that, as the parameter q is increased, the number of solution frequencies occasionally doubles, or suddenly a chaotic state may appear as a vertical band of frequencies. Then, the system may just as suddenly drop back into a simple state having only a few frequencies or perhaps a single frequency. At this specific operating condition, the pendulum system has a pure periodic motion, and the Poincaré section becomes a single point again. This circumstance was introduced earlier in Fig. 6.7, where the response of the Duffing equation to varying amplitude produced subharmonics. Figure 6.7 therefore is seen as a bifurcation diagram for this system.

In actuality, Fig. 7.22 is only a simple two-dimensional projection of the total parameter space of the pendulum example. The parameter q is but one of the pendulum's properties that must be studied. The true bifurcation diagram for the pendulum would be a four-dimensional plot of frequency ω as a function of q, g, and ω_D.

7.7 SUMMARY COMMENTS

The material presented in this chapter is merely a brief overview of the various devices used to assess the stability of a modeled system. The conditions for chaotic behavior are presented, along with various methods to test stability and to gain further insight into the regularity (or lack thereof) of the system.

7.8 PROBLEMS

7.1 Sketch the phase diagrams for the linear systems described by the following differential equations:

$$\dot{x} = x + y, \qquad \dot{y} = -5x - 3y,$$
$$\dot{x} = x - y, \qquad \dot{y} = x - 4y,$$
$$\dot{x} = x - 2y, \qquad \dot{y} = x + y,$$
$$\dot{x} = x - y, \qquad \dot{y} = -2x - 4y,$$
$$\dot{x} = 2x + 2y, \qquad \dot{y} = x + 4y,$$
$$\dot{x} = x - 2y, \qquad \dot{y} = 3x - 4y,$$
$$\dot{x} = 3x - y, \qquad \dot{y} = x + y,$$
$$\dot{x} = -x - 5y, \qquad \dot{y} = x + 3y,$$
$$\dot{x} = -x + y, \qquad \dot{y} = 2x + y.$$

7.2 Determine the singular points for the following systems. Classify and sketch their phase diagrams. You may find it useful to employ the method of isoclines in various regions of the phase space to assist you in this task:

$$\dot{x} = \sin y, \qquad \dot{y} = -\sin x,$$
$$\dot{x} = y e^y, \qquad \dot{y} = 1 - x^2,$$
$$\dot{x} = 1 - xy, \qquad \dot{y} = (x - 1)y,$$
$$\dot{x} = x - y, \qquad \dot{y} = 1 - xy,$$
$$\dot{x} = y\sqrt{1 - x^2}, \qquad \dot{y} = x\sqrt{1 - y^2}, \quad -1 \le x \le +1,$$
$$\dot{x} = -x + x^3, \qquad \dot{y} = y,$$
$$\dot{x} = -y - x + x^3, \qquad \dot{y} = y,$$
$$\dot{x} = -x - x^3, \qquad \dot{y} = y,$$
$$\dot{x} = \sin x (2 \cos x - 1), \qquad \dot{y} = y.$$

7.3 Show that all points along the line $y = cx/d$ are equilibrium points for the system

$$\dot{x} = ax + by, \qquad \dot{x} = cx + dy,$$

subject to the condition that $ad - bc = 0$.

7.4 Determine the location and type for all singular points of the differential equation

$$\frac{d^2 y}{dt^2} + a^2 y^2 = c^2, \qquad a, c = \text{real}.$$

If the phase-plane solution contains a separatrix, find the equation for it. Sketch the behavior of the system in a $y - dy/dt$ phase plane.

7.5 Locate all singular points for each of the following differential equations. Classify each singularity and thus determine the nature of the solution trajectories near each singularity.

(1) Simple pendulum:

$$\frac{d^2 \theta}{dt^2} + \frac{g}{L} \sin \theta = 0.$$

(2) van der Pol equation:
$$\frac{d^2x}{dt^2} - \mu(1-x^2)\frac{dx}{dt} + x = 0.$$

(3) Mass on a hard spring:
$$\frac{d^2x}{dt^2} + \omega_0^2(1+a^2x^2)x = 0.$$

(4) Mass on a soft spring:
$$\frac{d^2x}{dt^2} + \omega_0^2(1-b^2x^2)x = 0.$$

7.6 Determine the location and type of all singular points of the differential equation
$$\frac{d^2y}{dt^2} + \frac{dy}{dt} - 20y + 5y^3 = 0.$$

Sketch the solution trajectories for this system in a $y - dy/dt$ phase plane.

7.7 Show that the origin of the following system is classified as a FOCUS, but this changes to a CENTER when the system is linearized:
$$\dot{x} = -y - x\sqrt{x^2+y^2}, \qquad \dot{y} = x - y\sqrt{x^2+y^2}.$$

7.8 Determine the equilibrium points of the system of equations describing the one-dimensional (x) steady flow of a viscous gas subject to heat transfer given as
$$\frac{dv}{dx} = \kappa_1\sqrt{v}[2v - \sqrt{2v} + \theta],$$
$$\frac{d\theta}{dx} = \kappa_2\sqrt{v}\left[-v + \sqrt{2v} + \frac{\theta}{\gamma - 1} - c\right],$$

where v is the kinetic energy, θ is a measure of the temperature, and κ_1, κ_2, and c are constants.

7.9 Consider the population model for males, M, and females, F, given by
$$\dot{M} = -\alpha M + \gamma B(M)F,$$
$$\dot{F} = -\alpha F + \beta B(M)F,$$

where $\alpha > 0$, $\beta > 0$, and $\gamma > 0$. The death rates of both sexes are assumed to be the same, α. The birth rates of each sex are determined by the coefficient function $B(M) = 1 - e^{-kM}$, where $k > 0$. Note that when many males are available, the birth rate of males and females becomes γF and βF, respectively.

(1) If $\beta > \alpha$, show that the two equiulibrium states are at $(0, 0)$ and $(R(\beta/\gamma), R(1))$, where
$$R(z) = -\frac{z}{k}\log\left(\frac{\beta - \alpha}{\beta}\right).$$

(2) Linearize the population equations and classify the two equilibrium points.

(3) Validate that the function $M = \gamma F/\beta$ is a particular solution to the governing equations.
(4) Sketch the phase-plane solution.
(5) Discuss the stability of the population model.

7.10 Consider the two populations of species, S_1, S_1, which share a common food supply. The populations are described by the system of equations

$$\frac{dS_1}{dt} = S_1[b_1 - d_1(fS_1 + gS_2)],$$

$$\frac{dS_2}{dt} = S_2[b_2 - d_2(fS_1 + gS_2)].$$

(1) Classify the equilibrium points for the system.
(2) Provided $b_1 d_2 > b_2 d_1$, show that species S_2 dies out and species S_1 reaches a limiting value.

7.11 Examine the equations given below for stability by the second method of Lyapunov. In each case, use $V = x^2/2$ as a trial V function.
(1) $\dot{x} = x(x-1)$.
(2) $\dot{x} + x^2 = 0$.
(3) $\dot{x} + x^3 = 0$.
(4) $\dot{x} + x^2 = A$ $(A > 0)$.
(5) $\dot{x} + x^3 = A^3$ $(A > 0)$.

In the fourth and fifth equations above, arrange things so that the singular point of the differential equation lies at the origin. If you use a transformation $y = ax + b$ for this purpose, then use $V = y^2/2$ for your V function.

7.12 Consider the simple model of an economy in terms of income and expenditure, as given by the equations

$$\frac{dI}{dt} = I - aE,$$

$$\frac{dE}{dt} = b(I - E - T),$$

where I = gross individual revenue; E = rate of individual expenditures; T = rate of governmental taxation (and spending); a = factor indicating inflation, other losses; and b = factor indicating faith in the economy. The constants a, b are subject to the constraints $1 < \alpha < \infty$ and $1 \leq \beta < \infty$.

(1) If $T = T_0$ = constant, show that the solution has an equilibrium point.
(2) For $\beta = 1$, classify the equilibrium state and demonstrate that the system is oscillatory in nature.
(3) Consider the case in which governmental spending is held proportional to gross individual income, $T = T_0 + kI$, where $k > 0$:
   (a) If $k \geq (\alpha - 1)/\alpha$, determine whether an equilibrium state exists.
   (b) Describe how the economy will behave in this situation.

7.13 A small body of mass m_s occupies an orbit between a planet of mass M_p and a moon of mass M_m. The planet and its moon are at a distance of R apart. The force of gravitational attraction, F, between two bodies of masses m_1, m_2 and separated by a distance r is modeled as

$$F = -\frac{\gamma m_1 m_2}{r^2}.$$

(1) Derive the following equation of motion for the mass m:

$$\ddot{y} = \gamma \left[\frac{M_m}{(y-R)^2} - \frac{M_p}{y^2} \right],$$

where y is the orbit radius of m from the planet.

(2) Show that the equilibrium point for the linearized system is unstable.

7.14 This problem involves an analysis of the stability of bunch motion in an alternating gradient synchrotron. The governing equations for the system are

$$\frac{d\phi}{dt} = a\phi + bx + A,$$
$$\frac{dx}{dt} = c\sin\phi + dx + B,$$

where ϕ = phase of accelerating voltage, radians; x = radial deviation of orbit, cm; and a, b, c, d, A, and B are all constants.

(1) Determine the location and classifications of the singular points.
(2) Sketch the solution on the $x - \phi$ phase plane using the method of isoclines.
(3) Solve this set of equations using a numerical method.
(4) Graphically portray the results on the $x - \phi$ phase-plane.

For all of the above, use the following values:

$a = -4 \times 10^3 \text{ s}^{-1},$
$b = -2 \times 10^3 \text{ cm}^{-1} \text{ s}^{-1},$
$c = 1.5 \times 10^3 \text{ cm-s}^{-1},$
$d = -4.7 \text{ s}^{-1},$
$A = 4.6 \times 10^3 \text{ s}^{-1},$
$B = -8.2 \times 10^2 \text{ cm-s}^{-1}.$

Plot all of your phase plane results as x versus ϕ over the range $-4 \leq x \leq +4$, and $-\pi \leq \phi \leq \pi$.

7.15 Create a mathematical model of a garage door, as shown in Fig. 7.23. The lower guide is constrained to move vertically whereas the upper guide is constrained to move horizontally. In addition, the lower guide has a ball-bearing follower and the vertical component of friction can be neglected. The variables you are to include in

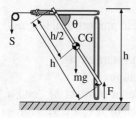

Figure 7.23. Schematic of a manually operated garage door.

your model are

θ = angle of door with respect to the horizontal plane, deg;
m = mass of garage door, lbm;
h = height of garage door, ft;
S = counterbalance force, lbf;
F = user-applied force, lbf;
L_x = lower x-component of reaction force, lbf;
U_x, U_y = upper x, y-components of reaction force, lbf.

(1) If the upper guide has a roller bearing such that $U_x = 0$, show that the system is stable for a specific design condition of the counterbalance force, S (constant).
(2) If the upper guide is instead designed using a sliding bearing such that $|U_x/U_y| = \mu$, how does this change the design of the counterbalance force, S, so that the system is in equilibrium in the open and closed positions?

7.16 One form of Volterra's equations is

$$\left. \begin{array}{l} \dot{x} = K_1 x - K_3 xy \\ \dot{y} = -K_2 y + K_4 xy \end{array} \right\} \quad K_1, K_2, K_3, K_4 > 0.$$

(1) For the general case (K_i as symbols), locate and classify the singularities.
(2) For the special case,

$$K_1 = 2, \quad K_2 = 8, \quad K_3 = 1, \quad K_4 = 2,$$

sketch the solution trajectories in the x-y plane.

CHAPTER 8

Case Studies

The following case studies have been chosen to demonstrate modeling principles and to provide examples of nonlinear problems and their solutions. As is usual in all models, it is possible to enhance these models and to make them much more detailed and accurate. One of the rules of modeling is to start simple, and this rule has been followed here. It is expected that the reader might wish to enhance his modeling skills by extending the models presented, since the essential behavior is already known.

8.1 LIGHTBULB MODEL

A simple lightbulb, shown diagrammatically in Fig. 8.1, is to be modeled. In this example, the purpose of the mathematical model is to discern the time rate of change of the surface temperature of the glass bulb. It is important to keep the model simple at first, but one also must attempt to understand the implicit assumptions that often are invoked unconsciously in the simplification process. As an example, consider the simplified model shown in Fig. 8.1. The assumptions inherent in this greatly simplified model are:

1. The surroundings are at a constant temperature of T_0. In addition, the glass is assumed to initially be at the temperature T_0 and its temperature remains uniform.
2. The glass bulb is assumed to be spherical with radius R.
3. The bulb has a constant specific heat, C_p, density, ρ, absorptivity, α, and thickness, δ.
4. The glass is a thin shell with $h \ll R$; thus, the volume of the shell is $4\pi R^2 \delta$.
5. Neglect the conduction energy loss to the base and the convection loss to the surrounding air.
6. The glass radiates energy as a black body.
7. The fraction of the input power, P, that is absorbed by the glass is instantaneously equal to αP.

8.1.1 Method 1: Linearization

The energy balance for a lightbulb in word form can be written

[Rate of Energy Gain] = [Radiant Energy In] − [Radiant Loss]
                           − [Convective Loss] − [Conduction Loss].

This same equation may be written in symbolic form, and if conduction and convection

8.1. Lightbulb Model

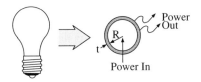

Figure 8.1. Simplified lightbulb model.

are neglected, it becomes

$$\rho C_p (4\pi R^2 \delta) \frac{dT}{dt} = \alpha P - \sigma A_g \left(T^4 - T_0^4 \right). \tag{8.1}$$

If the nonlinear term, T^4, is removed by linearization about $T = T_0$ and two new variables are defined,

$$y = T - T_0 \quad \text{and} \quad x = \delta t \left/ \frac{mC}{4 A_g \sigma T_0^3} \right.,$$

Eq. (8.1) reduces to

$$\frac{dy}{dx} + y = F, \qquad y(0) = 0, \tag{8.2}$$

where $F = \alpha P / (4 \sigma A_g T_0^3)$. This equation has a solution

$$y = F(1 - e^{-x}). \tag{8.3}$$

This result is seen as having a single parameter, F, and is thus relatively parameter independent. The analysis thus provides the designer with only one parameter which may be varied to affect the final solution. A plot of the solution mapped to the case in which the ratio of initial to final temperatures is 0.25 is shown in Fig. 8.2. To scale back to the dimensional world, however, requires the two parameters τ and F. This plot does not vary with the parameters of the system since the abscissa is a dimensionless time and the ordinate is nondimensional as well. This independence of initial conditions and parameter values can be demonstrated *only* in linear systems. This is one of the differences in properties between nonlinear and linear systems, as discussed in Chapter 2, Section 2.1.

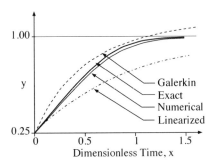

Figure 8.2. Comparison of the linearized, exact, and Runge-Kutta solutions.

8.1.2 Method 2: Exact Solution of Nonlinear Problem

In this method, the choice of transformation used is different and simpler than that used in the linearization model, above. The differential equation, Eq. (8.1), is transformed using $y = T/T_m$, and $x = t/\tau$. This results in the transformed system

$$\frac{dy}{dx} + \tau \left[\frac{\sigma A_g T_m^3}{4\pi \rho C_p R^2 \delta} \right] y^4 = \tau \left[\frac{\alpha P + \sigma A_g T_0^4}{4\pi \rho C_p R^2 \delta} \right]. \tag{8.4}$$

By setting the two coefficients to be one, the two transformation variables are determined, and thus the temporal and thermal scales for the problem, as

$$\tau = \frac{4\pi \rho C_p R^2 \delta}{\sigma A_g T_m^3}, \tag{8.5}$$

$$T_m^4 = T_0^4 + \frac{\alpha P}{\sigma A}. \tag{8.6}$$

In Eq. (8.6), T_m is seen readily as the maximum steady-state temperature where dy/dx vanishes. Making these substitutions, the differential equation reduces to the simple equation

$$\frac{dy}{dx} + y^4 = 1, \qquad y(0) = y_0 = \frac{T_0}{T_m}. \tag{8.7}$$

This equation is directly integrable and the solution is

$$x = \frac{1}{2}[\tan^{-1}(y) - \tan^{-1}(y_0)] + \frac{1}{4} \ln\left[\frac{(1+y)(1-y_0)}{(1-y)(1+y_0)}\right]. \tag{8.8}$$

This solution is not of the usual form (y as a function of x). A plot of the exact solution for $y_0 = 0.25$ is shown in Fig. 8.2.

8.1.3 Method 3: Galerkin Solution

The solution of Eq. (8.7) using the Galerkin method is predicated on our knowledge about the character of the solution. By looking at Eq. (8.7), it is apparent that the response, y, starts at an initial condition of $y(0) = y_0$ and grows as a decaying exponential, approaching a value of $y(\infty) = 1$ when the derivative vanishes. This is confirmed by examining the exact solution shown in Fig. 8.2. An obvious choice for a Galerkin approximation function is thus

$$\tilde{y} = A + Be^{-x}. \tag{8.9}$$

If one applies the initial condition $y(0) = 0.25$, this yields a residual function, ϵ, of

$$\epsilon = 1 + \frac{(1-4A)}{4}e^{-x} - \left[A + \frac{(1-4A)}{4}e^{-x}\right]^4. \tag{8.10}$$

The Galerkin normalization condition is thus

$$\int_{x=0}^{1} \epsilon e^{-x}\, dx = 0, \tag{8.11}$$

which yields the value $A = 1.375$, and the solution is then

$$\tilde{y} = 1.375 - 1.125e^{-x}. \tag{8.12}$$

It is left as an exercise to the reader to consider if the limits of the integral in Eq. (8.11) are the best choice. These limits determine the region over which the error associated with the assumed solution is minimized.

8.1.4 Method 4: Numerical Solution

The numerical solution of Eq. (8.7) is relatively straightforward. A Runge-Kutta solution is depicted in Fig. 8.2. Since the range of variation of x is on the order of one, a step size of $\Delta x = 0.01$ is used. Greater numerical accuracies are possible using a smaller value, but the character of the solution is captured readily at this step size.

8.1.5 Summary Comments

The solution of this simple mathematical model provides insight into the accuracy and limitations of the three methods presented. The linearized model has the worst overall performance, yet it also provides an extremely simple and useful analytical result. Its short-time behavior is acceptable and the asymptotic solution is correct, but the behavior of the solution in mid-domain is not acceptable. In ascertaining the influence of design parameters on the lightbulb model, this is the most direct and straightforward solution. The use of a symbolic math program made it possible to obtain an exact solution in a very short time, but the result obtained was inverted and in the form of $x = f(y)$ rather than $y = g(x)$. The solution function, Eq. (8.8), was sufficiently complex to render it nonreversible unless a power-series approach was invoked. The Galerkin solution does well over the range in which it was optimized ($x[0..1]$), but for longer times, y grows substantially and eventually reaches unacceptable levels. The asymptotic value of this solution is $y = 1.375$. Note that this is not a general solution, being strictly valid only for a specific initial condition. This property is shared by the numerical solution presented as Method 4.

8.2 AUTOMOBILE CARBURETOR MODEL

Consider an automobile carburetor system as depicted in Fig. 8.3. The fuel and air delivery systems are to be modeled assuming that the input variables are the throttle angle, α, and the engine speed, N. To assist in formulating the approach of this model, the observations of Table 8.1 are given. Because of the difficulties associated with mathematical modeling of the air/fuel flow in the intake system with a variable turbulence-related loss at the butterfly throttle valve, a simpler, heuristically determined model is proposed and developed here. The data indicate that there is the possibility that a simple mathematical expression exists

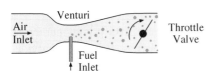

Figure 8.3. Simplified diagram of an automobile carburetor.

TABLE 8.1. Manifold Pressure Changes with Engine Speed, Throttle Angle

Engine speed (rpm)	Manifold pressure (in. Hg)	Throttle angle (deg)
N_{idle}	−25	0
N_{max}	−28	0
N_{idle}	0	90
N_{max}	−10	90

that describes the manifold pressure as a function of throttle angle, α, and engine speed, N. The following function is presented as having the desired character for the manifold pressure, P_{man}, as a function of these variables and the idle and maximum engine speeds, N_{idle} and N_{max}, respectively, and the idle throttle angle, α_0:

$$\frac{P_{\text{man}}}{P_a} = 1 + C_\alpha \cos(\alpha) + C_N \left(\frac{N - N_{\text{idle}}}{N_{\text{max}} - N_{\text{idle}}} \right). \tag{8.13}$$

This assumed functional relationship for the manifold pressure is harmonic with respect to the throttle position and linear with respect to the engine speed. The formulation is written as a dimensionless relationship to make it more general in nature and less specific to this example. The influence coefficients, C_α, C_N, describe the sensitivity of the manifold pressure to the throttle angle and engine speed, respectively, and must be determined by matching Eq. (8.13) to Table 8.1. This results in the values $C_\alpha = -0.8333$ and $C_N = -0.1$, producing the formula

$$\frac{P_{\text{man}}}{P_a} = 1 - 0.8333 \cos(\alpha) - 0.1 \left(\frac{N - N_{\text{idle}}}{N_{\text{max}} - N_{\text{idle}}} \right). \tag{8.14}$$

The volume flow rate of the air through the engine, $\dot{Q}$, can be written

$$\dot{Q} = \frac{Q_0 N}{2}, \tag{8.15}$$

where Q_0 is the volumetric exchange of the engine every two revolutions (for a four-cycle engine). Because of the losses in the system, the temperature change of the air flowing into the manifold is assumed to be negligible, or $T_{\text{man}} = T_0$. The resulting mass flow rate of the air, $\dot{m}_{\text{air}}$, then is

$$\dot{m}_{\text{air}} = \rho \dot{Q} = \frac{P_{\text{man}}}{RT_0} \frac{Q_0 N}{2}. \tag{8.16}$$

The total fuel flow rate is assumed to be properly mixed by the carburetor such that the fuel/air ratio, $\mathcal{FA}$, is approximately constant. Thus, the fuel delivered is

$$\dot{m}_{\text{fuel}} = \mathcal{FA} \, \dot{m}_{\text{air}}. \tag{8.17}$$

The data in Table 8.2 are for a typical engine and fuel. These data permit numerical computations of the fuel delivery rate. The value of this simple model is that it provides the right qualitative relationship between the mass flow rates of fuel and air with respect to both engine speed and throttle setting. The resulting fuel delivery rate at varying throttle

8.3. Automobile Engine Model

TABLE 8.2. Property Data for Air and the Fuel Flow System

Variable	Value	Units
Air properties		
Specific heat ratio, k	1.40	
Gas constant (air), R_a	53.35	ft lbf/lbm-°R
Atmospheric pressure, P_a	14.70	psia
Air temperature, T_a	70.00	°F
Carburetor parameters		
Carburetor diameter, D^*	1.00	in.
Air/fuel ratio, $\mathcal{AF}$	13.60	lbm-air/lbm-fuel
Engine properties		
Idle speed, N_{idle}	900	rpm
Max. engine speed, N_{max}	6000	rpm
Engine displacement, Q_0	300	in.^{3}

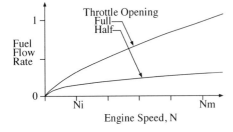

Figure 8.4. Normalized fuel flow at varying carburetor throttle angles.

positions and at varying engine speeds is shown in Fig. 8.4. This simple model is also usable as a driver for a thermodynamic model of an automobile engine. This relationship is given by

$$\dot{m}_{\text{fuel}} = \mathcal{FA}\dot{m}_{\text{air}} = \mathcal{FA}\frac{P_a\left[1 + C_\alpha \cos(\alpha) + C_N\left(\frac{N-N_{\text{idle}}}{N_{\text{max}}-N_{\text{idle}}}\right)\right]}{RT_0}\frac{Q_0 N}{2}. \qquad (8.18)$$

It is noteworthy that these curves are concave with respect to engine speed and that the maximum fuel delivered drops rapidly with a reduction from full throttle. At very low throttle angles, the fuel flow rate peaks early and then falls with increasing engine speed. This results in a dramatic reduction in the available energy delivered to the engine. Since frictional losses also rise with increasing engine speed, the system is self-limiting.

8.3 AUTOMOBILE ENGINE MODEL

Consider the design of a mathematical model of an automobile engine shown in Fig. 8.5 with the following properties:

- The fuel and air delivery systems are assumed to be a function of the throttle angle, α.
- The engine is a rotating inertial system with a variable load torque, T_L.
- The combustion process converts the fuel–air mixture to useful power. Assume that the potential fuel energy can be converted to mechanical energy (before bearing and friction losses) at a constant efficiency, $\eta_{\text{th}} = 0.45$.

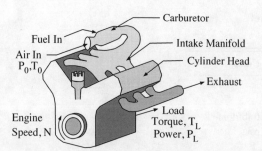

Figure 8.5. Simplified diagram of an automobile engine.

- The dynamic (not static) bearing losses and the piston-to-cylinder frictional losses are to be modeled as proportional to N^2.

The model is to be scaled to represent a six-cylinder engine capable of producing a net power of 140 horsepower (hp) at 6000 revolutions per minute (rpm). Note also that the engine will idle at 750 rpm when the throttle angle is at α_0 deg and there is no load on the engine.

The model's performance is to be demonstrated in the following manner:

1. Assume that the engine is loaded to steady state at varying rotational speeds. Graphically depict the maximum output horsepower, P_L, and torque, T_L, vs engine rotational speed, N.
2. Solve for the engine speed vs time if there is no load on the engine ($T_L = 0$) and the throttle is snapped from idle setting to full throttle and held there. Depict this result graphically over the range $N(\alpha_0, 8000)$ rpm. Also graph the variation of power delivered by the fuel vs time and the power losses over time.

8.3.1 Model Development

Consider that the engine has C cylinders and a rotating inertia of I. The torque developed from the combustion in the cylinders, T_{comb}, the frictional losses, T_{fric}, and the external load torque, T_{load}, combine to accelerate the engine as given by

$$T_{\text{comb}} - T_{\text{fric}} - T_{\text{load}} = I\dot{N}. \tag{8.19}$$

The frictional torque can be simply represented as a function of engine speed,

$$T_{\text{fric}} = C_f \dot{\theta}, \qquad \dot{\theta} > 1, \tag{8.20}$$

where C_f is a frictional coefficient. The combustion-generated torque also is modeled in a simplistic manner, using the fuel flow rate, $\dot{m}_{\text{fuel}}$, and a thermal efficiency, η_{th}, as the governing variables affecting the torque:

$$T_{\text{comb}} = \eta_{\text{th}} \dot{m}_{\text{fuel}} \delta H_c.$$

The relationship between fuel flow rate, throttle angle, and engine speed is

$$\dot{m}_{\text{fuel}} = \mathcal{AF} \frac{P_0 \left[1 + C_\alpha \cos(\alpha) + C_N \frac{N - N_{\text{idle}}}{N_{\text{max}} - N_{\text{idle}}}\right]}{R_a T_0} \frac{Q_0 N}{2}. \tag{8.21}$$

8.4. Surge Analysis for a Hydroelectric Power Plant

TABLE 8.3. Relevant Data for the Engine, Air, and Fuel Flow Systems

Variable and variable definition	Value	Units
Air properties		
Specific heat ratio, k	1.40	
Gas constant (air), R_a	53.35	ft-lbf/lbm-°R
Atmospheric pressure, P_a	14.70	psia
Air temperature, T_a	70.00	°F
Fuel flow rate		
Carburetor diameter, D^*	1.00	in.
Air/fuel ratio, $\mathcal{AF}$	13.60	
Combustion process		
Thermal efficiency, η_{th}	0.46	
Heat of combustion, δH_c	20,000	Btu/lbm
Engine properties		
Engine inertia, I	0.25	ft-lbf-s²
Frictional loss coefficient, C_f		ft-lbf-s²/rad²
Idle throttle angle, α_0		deg
Engine speed, N		rpm
Idle speed, N_{idle}	900	rpm
Max. engine speed, $N_{\max}$	6000	rpm

Note that the functional for the fuel flow rate is dependent on two inputs: throttle angle and the square of the engine speed.

8.3.2 Experimental Data

Data for the proposed engine are given in Table 8.3. In Eq. (8.21), the influence coefficients are $C_\alpha = -0.8333$ and $C_N = -0.1$. The engine displacement, Q_0, is the volume flow rate through the engine every two revolutions (for a four-cycle engine).

8.4 SURGE ANALYSIS FOR A HYDROELECTRIC POWER PLANT

The physical situation shown in Fig. 8.6 delivers water from a large reservoir of constant elevation, H, to a hydroelectric turbine. A surge tank of diameter, D, is included to prevent excessive pressure rises in the pipe whenever the valve is closed quickly during an emergency. The following problem is a design study related to the behavior of the system during such an emergency.

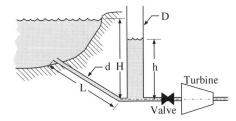

Figure 8.6. Diagram of the hydroelectric power plant. (Note that the water height of the reservoir is essentially constant.)

TABLE 8.4. Physical-Property Data for the System

Variable	Value	Units
g	32.2	ft/s^2
H	100	ft
h_0	88	ft
f	0.024	
L	2000	ft
d	2	ft
t_c	6	s
k	21.4	ft$^{5/2}$/s
D	6	ft

8.4.1 Development of the Model

By assuming the density to be constant and neglecting the effect of acceleration of water in the surge tank, a momentum balance on the water in the inclined pipe leads to

$$L \frac{du}{dt} = g(H - h) - \frac{f}{2} \frac{L}{d} \frac{u|u|}{2}. \tag{8.22}$$

Here, h = height of the water in the surge tank (h_0 under steady flow conditions), g = local acceleration of gravity, f = Moody friction factor, u = mean velocity in the inclined pipe, and t = time. Also, the conservation of mass (continuity) requires that

$$\frac{\pi d^2}{4} u = \frac{\pi D^2}{4} \frac{dh}{dt} + Q_v. \tag{8.23}$$

The flow rate, Q_v, through the valve during the time period of the valve closure ($0 < t < t_c$) is approximated by a time-linear closing relation,

$$Q_v = k \left(1 - \frac{t}{t_c}\right) \sqrt{h - h'}, \tag{8.24}$$

where the constant k is specific to the large flow control valves used in the water conduit. The pressure upstream of the turbine (turbine head), h', depends on the particular operating load at the turbine during the emergency shutdown. It also is assumed to be a constant for the purposes of this analysis.

Typical data are given in Table 8.4 allowing an analysis of the variation of the level in the surge tank with time. The initial head height in the surge tank prior to initiating the closing of the valve is set at its original steady-state value, $h_0 - (Q_{v0}/k)^2$, where Q_{v0} is the original steady flow rate:

$$L \frac{du}{dt} = g(H - h) - \frac{f}{2} \frac{L}{d} u|u|, \tag{8.25}$$

$$\frac{\pi d^2}{4} u = \frac{\pi D^2}{4} \frac{dh}{dt} + Q_v, \tag{8.26}$$

$$Q_v = k \left(1 - \frac{t}{t_c}\right) \sqrt{h - h'}, \tag{8.27}$$

where h = height of the water in the surge tank (h_0 under steady flow conditions),

8.4. Surge Analysis for a Hydroelectric Power Plant

h' = turbine head, f = Moody friction factor, u = mean velocity in the pipe, g = local acceleration of gravity, t = time, and Q_v = valve flow rate during $0 < t < t_c$. Rewriting the given differential equations,

$$\frac{du}{dt} = \frac{g}{L}(H-h) - \frac{f}{2d} u|u|, \tag{8.28}$$

$$\frac{dh}{dt} = \left(\frac{\pi d^2}{4} u - Q_v\right) \frac{4}{\pi D^2}. \tag{8.29}$$

The equations are coupled in that both dependent variables, u and h, appear in each equation. This system of equations must be solved simultaneously.

8.4.2 Numerical Integration Scheme

The Runge-Kutta method is used to solve the two first-order differential equations, (8.28) and (8.29), simultaneously. Among many numerical methods available for solving initial-value problems, the Runge-Kutta method is probably the one most frequently used on computers. To solve this problem, a set of initial conditions must be specified. To initiate this problem, the condition $h_0 = 80$ ft at $t = 0$ is observed. To find an initial value for u_0 at $t = 0$ requires the solution of the equation $du/dt = 0$ with $h = h_0$. Hence, letting

$$\frac{g}{L}(H - h_0) - \frac{f}{2d} u_0 |u_0| = 0 \tag{8.30}$$

yields

$$u_0 = \sqrt{\frac{2gd}{Lf}(H - h_0)}. \tag{8.31}$$

The two initial conditions are, at $t = 0$,

$$h(0) = 80, \tag{8.32}$$

$$u(0) = u_0 = \sqrt{\frac{2gd}{Lf}(H - h_0)}. \tag{8.33}$$

Now, with Eqs. (8.27), (8.30), and (8.31) and the initial conditions, Eqs. (8.32) and (8.33), the problem is to determine the values of h and u as time increases by the increment Δt. The original steady flow rate, Q_{v0}, is expressed as

$$Q_{v0} = \frac{\pi d^2}{4} u_0.$$

Then, $Q_{v0} = 51.783$. Also, two extreme values for h' are to be studied. The specific values of h' depend on the particular emergency at the turbine. Its original steady value, $h' = h_0 - (Q_{v0}/k)^2$, is numerically 74.145. The other value is $h' = 0$.

8.4.3 Results

The result of the numerical analysis of the head height following the emergency valve closure is depicted in Fig. 8.7. This is specific to the parameters given in Table 8.4. As

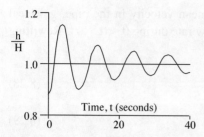

Figure 8.7. Transient response of a hydroelectric power plant.

the valve closes during the time $0 < t < t_c$, the height of the surge tank increases steadily until the valve closure is completed and $h' = 74.145$. The initial oscillations in the surge tank slowly die out due to frictional losses in the system. A very interesting result occurs, however, during the valve transient period if the turbine inlet pressure is fixed at $h' = 0$. The reader may wish to repeat this analysis for a variety of conditions.

When the valve is closed completely, the flow rate through the valve, Q_v, is zero and the two head heights are identical ($h = H$). However, because of the inertia of the water through the inclined pipe, water will continue to flow through the pipe and into the surge tank. This flow becomes oscillatory and the character of the oscillation depends highly on the diameter of the surge tank as well as the turbine head, h'. This can be observed by comparing the attached plots for the head, h.

Finally, it is observed that the diameter of the surge tank has a dramatic influence on the damping coefficient. When the diameter is small, the oscillation is underdamped. Larger diameters, however, increase the damping and the flow oscillations become overdamped.

8.5 CONSTANT-DECELERATION SHOCK ABSORBER

Consider the design of a constant-deceleration hydraulic shock absorber using a (theoretically) incompressible fluid to dissipate the energy applied to the shock absorber. The basic model is shown Fig. 8.8.

Using this simple design, the governing equations are developed and their validity is analyzed for possible application to a real-world design for such a shock absorber. In this system, an incompressible fluid (oil, for example) is forced out of a discharge orifice of constant area during the energy-absorbing stroke. The resulting deceleration is not constant, but the equations that describe the motion for this problem give a basis from which the constant-deceleration situation can be more nearly approximated.

The basic force equation is quite simple:

$$F = pA, \tag{8.34}$$

where F = the decelerating force, p = the fluid pressure, and A = the piston area. The

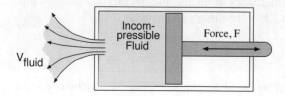

Figure 8.8. Piston pushing an incompressible fluid through an orifice.

8.5. Constant-Deceleration Shock Absorber

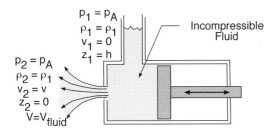

Figure 8.9. Piston with a fluid head reservoir.

problem is to express the fluid pressure, p, in terms of appropriate parameters of the design, such as velocity of the piston and discharge orifice area.

For an incompressible fluid, the volume displaced as the piston moves down the cylinder will, of course, equal the volume discharged through the orifice:

$$Q_p = Q_o,$$

or, for a given increment of time Δt,

$$A \frac{dx}{dt} \Delta t = (ca) v \Delta t,$$

where dx/dt = piston velocity, a = orifice area, c = discharge coefficient, and v = fluid velocity. This simplifies to

$$A \frac{dx}{dt} = cva. \tag{8.35}$$

Now, by applying Bernoulli's equation (see Fig. 8.9), we obtain an expression for fluid velocity,

$$v = \sqrt{2gh}, \tag{8.36}$$

where v = fluid velocity, h = fluid head, and g_c = gravitational acceleration constant. The fluid head is directly proportional to the pressure drop across the orifice,

$$h = p/\rho, \tag{8.37}$$

where ρ is the fluid density.

Now, combining Eqs. (8.35), (8.36), and (8.37) and solving for pressure, p, we get

$$p = \left(\frac{A}{ca\sqrt{2gk}} \right)^2 \left(\frac{dx}{dt} \right)^2. \tag{8.38}$$

Substituting Eq. (8.38) into Eq. (8.34) yields

$$F = \frac{A^3}{(ca\sqrt{2gk})^2} \left(\frac{dx}{dt} \right)^2$$

or

$$F = \beta \left(\frac{dx}{dt} \right)^2, \tag{8.39}$$

where

$$\beta = \frac{A^3}{(ca\sqrt{2gk})^2}, \qquad (8.40)$$

8.5.1 Equations of Motion for Incompressible Flow

Using Newton's equation, we can begin to develop the dynamic equations for this model. In essence, it is simply an expression that states that the sum of the forces on the piston equal the product of the piston mass times its acceleration,

$$M\frac{d^2x}{dt^2} = \sum F,$$

or, neglecting all frictional effects between piston and cylinder, this becomes

$$M\frac{d^2x}{dt^2} = -F_{\text{piston}}. \qquad (8.41)$$

Substituting Eq. (8.39) into Eq. (8.41), we get

$$M\frac{d^2x}{dt^2} + \beta\left(\frac{dx}{dt}\right)^2 = 0, \qquad (8.42)$$

which is the equation of motion for the piston assembly.

A model of this sort is fairly straightforward, and much of its behavior can be predicted by deriving the characteristic equations of motion that describe acceleration, velocity, and position. In particular, let's look at velocity as a function of time, $v(t)$, position as a function of time, $x(t)$, velocity as a function of position, $v(x)$, and acceleration as a function of position, $a(x)$.

Recalling Eq. (8.42), let's make the substitution of $v = dx/dt$ and $dv/dt = d^2x/dt^2$, giving

$$M\frac{dv}{dt} + \beta v^2 = 0.$$

Separating variables,

$$M\frac{dv}{v^2} = -\beta\,dt,$$

and integrating this gives

$$M\left(\frac{-1}{v}\right) = -\beta t + c_1.$$

Assuming that, at $t = 0$, $v = v_1$,

$$c_1 = \frac{-M}{v_1},$$

8.5. Constant-Deceleration Shock Absorber

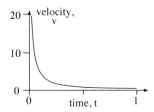

Figure 8.10. Graph of the velocity-time characteristic.

so that

$$v = \left[\frac{1}{v_1} + \frac{\beta t}{M}\right]^{-1}.$$

An analytical plot of velocity as a function of time (see Fig. 8.10) reveals an important behavior, namely, that velocity will *asymptotically* approach zero. That is, theoretically, the piston velocity will never completely reduce to zero. In other words, the fluid will not be able to completely halt the piston's travel; rather, it will only slow it.

By integrating the velocity-time characteristic derived above, we can derive the displacement-time characteristic.

$$\frac{dx}{dt} = \left[\frac{1}{v_1} + \frac{\beta t}{M}\right]^{-1},$$

$$dx = dt \left[\frac{1}{v_1} + \frac{\beta t}{M}\right]^{-1}.$$

Thus,

$$x = \frac{M}{\beta} \ln\left(\frac{1}{v_1} + \frac{\beta t}{M}\right) + c_2.$$

when $t = 0$, $x = 0$, and so, the constant of integration becomes

$$c_2 = -\frac{M}{\beta} \ln\left(\frac{1}{v_1}\right).$$

Substituting gives

$$x = \frac{M}{\beta} \left(\ln\left(\frac{1}{v_1} + \frac{\beta t}{M}\right) - \ln\left(\frac{1}{v_1}\right) \right)$$

or

$$x = \frac{M}{\beta} \ln\left(1 + \frac{\beta v_1 t}{M}\right), \tag{8.43}$$

which is the displacement-time characteristic equation. An analytical graph (Fig. 8.11) reveals that x versus t is *not* asymptotic, and therefore, we learn that the piston eventually will reach the end of the cylinder.

To investigate the velocity-displacement characteristic, we return to Eq. (8.42):

$$M\frac{d^2x}{dt^2} + \beta \left(\frac{dx}{dt}\right)^2 = 0.$$

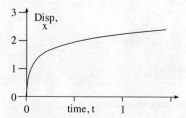

Figure 8.11. Graph of the displacement-time characteristic.

Now, letting $dx/dt = v$ and $d^2x/dt^2 = dv/dt = (dv/dx)(dx/dt) = v\, dv/dx$, we get

$$Mv\frac{dv}{dx} + \beta v^2 = 0.$$

Again, separating variables and integrating gives

$$M \ln v = -\beta x + c_3$$

and, at $x = 0$, $v = v_1$, so that $c_3 = M \ln v_1$ and the above equation becomes

$$v = v_1 e^{-\beta x/M}. \tag{8.44}$$

This is the velocity-displacement characteristic equation. Note that its behavior is similar to that shown in the velocity-time curve (see Fig. 8.10), in that both time and displacement, theoretically, must be infinite for the velocity to reach zero.

Finally, the acceleration-displacement characteristic is a very important relationship for study. In particular, the integral of this function reveals information about the work performed in reducing the kinetic energy of the load and piston assembly.

Using the velocity-displacement characteristic just derived, we substitute into Eq. (8.42),

$$M\frac{d^2x}{dt^2} = -\beta v^2$$

or

$$\frac{d^2x}{dt^2} = -\frac{\beta}{M}\left(v_1 e^{-\beta x/M}\right)^2.$$

Rewriting gives the acceleration as a function of position,

$$a = -\frac{\beta v_1^2}{M} e^{-2\beta x/M}.$$

An analytical graph of this function is given in Fig. 8.12; the area under the curve is proportional to the work necessary to reduce the kinetic energy of the piston. Note that

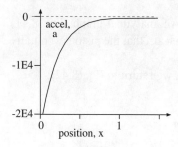

Figure 8.12. Graph of the acceleration-displacement characteristic.

8.5. Constant-Deceleration Shock Absorber

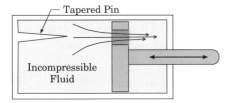

Figure 8.13. Tapered-pin approach to the design.

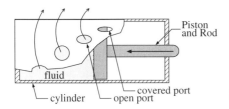

Figure 8.14. Multihole approach to the design.

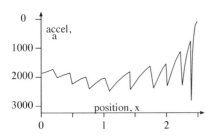

Figure 8.15. Multihole acceleration-displacement relationship.

the steeply sloping portion of the curve implies that large forces will be in play for short periods of time during the stroke of the piston.

8.5.2 Procedure for Constant-Deceleration Design

From the previous analysis, we see that a simple constant-area orifice will not yield a constant deceleration. Recall that β is proportional to the cube of the orifice area. From the acceleration-displacement characteristic, we see that if the orifice area diminishes as a function of x, the deceleration can be smoothed. One design approach might be to place the orifice on the piston head and have it descend upon a tapered pin, as in Fig. 8.13.

Returning to the previous formulation, we attempt a design that gives a deceleration that is within a tolerable range, even though it does not give a *constant* deceleration. This is accomplished by diminishing the discharge area in discrete steps. The piston wall is drilled with several holes such that, as the piston moves through its stroke, it covers discharge holes, thus reducing the total discharge area.

Figure 8.14 describes the basic design. All of the governing equations developed previously still hold. The effect is a piecewise solution to several such pistons, each with less total orifice area. The associated acceleration-displacement characteristic is depicted in Fig. 8.15.

Note that this plot resulted from an assumption that the hole closure is instantaneous. In effect, the holes are very thin slits cut around the surface of the cylinder. This is not very desirable since, depending on the specific geometry, the structural integrity of the cylinder might be compromised. This offers one of several avenues of further study, namely, to model various, more realistic, methods of hole closure. The simplest is to consider a linear hole closure. Computationally, this would be obtained by simply scaling the hole area down continuously as the piston moves across its position. This, however, would still model a square hole, and the construction cost of such a design would be higher.

8.6 FLUID MECHANICS OF BLOOD FLOW IN THE KIDNEY

The flow of blood through the kidney is shown in Fig. 8.16. If the kidney consisted only of passive, rigid-walled vessels, there would be a linear relationship between the flow rate of an incompressible fluid and the pressure difference across the system. As noted in Fig. 8.17, however, the flow of blood through the kidney is autoregulated, or somehow controlled by a mechanism intrinsic to the kidney. An experimental study of this behavior is reported by Rothe [37]. The principal goal here is to create an analytic model that attains the same character as that portrayed in Fig. 8.17, but which also provides insight into the autoregulatory mechanism. To accomplish this goal, we first gather some important information about the basic fluid flow system. By looking at the pressure drop throughout the fluid flow from arterial inlet to venous outlet (see Fig. 8.18), the most dramatic pressure drop is found to be at the lowest level in the arterial flow, or the arterioles. What, then, is so unusual or different about this segment of the flow system. As portrayed in Fig. 8.18, there are muscle fibers wrapped circumferentially around the vessel walls of the arterioles. This is extremely suggestive that there is an active wall effect in this region. When the kidney is denervated, it still autoregulates, inferring that some internal

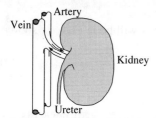

Figure 8.16. Blood flow into and out of the kidney.

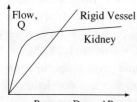

Figure 8.17. Autoregulated versus rigid vessel flow.

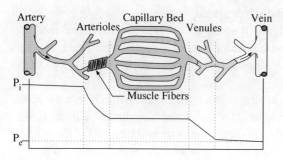

Figure 8.18. Symbolic drawing of the vascular bed.

8.6. Fluid Mechanics of Blood Flow in the Kidney

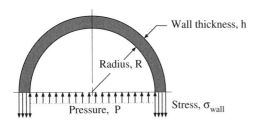

Figure 8.19. Force balance on the arterial vessel wall.

subsystem is responsible for this behavior. Added insight comes from a simple analysis of the interaction between alterations in the diameter of the vessel and the internal pressure. Figure 8.19 is a symbolic cross section of the arteriole at this level, showing all pressure and stress forces. By performing a force balance across the vessel, and assuming an incompressible wall, or $hR = H_0R_0$, we obtain a nonlinear relationship between the vessel radius, R, and the internal pressure, P:

$$R = \sqrt{\frac{h_0 R_0 \sigma_w}{P}}. \tag{8.45}$$

If we assume that the smooth muscle of the arterioles must function to hold the wall hoop stress, σ_w, at some constant level, then as the pressures increase, the vessel diameter must decrease, thereby elevating the viscous resistance to flow at higher pressures. This assumed mechanism appears to be acting in an appropriate manner.

Under the assumptions of a one-dimensional flow of a homogeneous, Newtonian fluid with constant properties flowing steadily in a vessel with negligible wall curvature and negligible body-force effects, the momentum equation for the flow in the arterioles is

$$\frac{dP}{dx} + \frac{8\mu Q}{\pi R^4} = 0 \tag{8.46}$$

and the continuity equation is simply

$$\frac{\partial R^2}{\partial t} = \frac{\partial Q}{\partial x} = 0. \tag{8.47}$$

Equation (8.47) therefore is stating that $R = R(x)$, and $Q = Q(t)$. Because of the assumption that inertial effects are neglected, the implication is that Q is not time dependent, but a constant. Combining Eqs. (8.45) and (8.46), and introducing the scaling variables $y = P/P_i$, $z = x/L$ yields

$$\frac{\pi (h_0 R_0 \sigma_w)^2}{8\mu Q \pi} \frac{dy}{dz} + y^2 = 0, \qquad y(0) = 1. \tag{8.48}$$

Equation (8.48) can be simplified by choosing a typical wall stress that forces the coefficient of the derivative term to unity, or

$$\sigma_w = \frac{1}{h_0 R_0} \sqrt{\frac{8\mu QLP_i}{\pi}}. \tag{8.49}$$

This reduces Eq. (8.48) to

$$\frac{dy}{dz} + y^2 = 0, \qquad y(0) = 1.$$

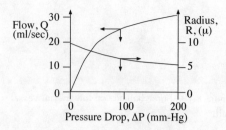

Figure 8.20. Flow rate and vessel radius versus perfusion pressure for $P_e = 50$ mm Hg.

8.6.1 Exact Solution

This nonlinear equation may be directly integrated to obtain the analytic solution function

$$y = \frac{1}{z+1}. \tag{8.50}$$

Note that, as the dimensionless position $z \to 1$, the dimensionless pressure drops to $y = 0.5$. If one recasts Eq. (8.50) back into the dimensional world and solves for the flow rate with $P = P_e$ at $z = 1$, this yields

$$Q = \frac{\pi (h_0 R_0 \sigma_w)^2}{8\mu L} \left[\frac{1}{P_e} - \frac{1}{P_i} \right]. \tag{8.51}$$

Equation (8.51) is the analytic solution for the steady-state flow rate, Q, in the arterioles, and its dependence on the driving arterial pressure, P_i, and capillary pressure, P_e. This result is plotted for typical physiological values in Fig. 8.20, holding $P_e = 50$ mm Hg, or approximately one-half of the normal arterial pressure. The comparison between the qualitative character portrayed in Fig. 8.20 and that in Fig. 8.17 demonstrates the feasibility of this model.

8.6.2 Galerkin Solution

In attempting to discern a Galerkin solution for this problem, if one examines the basic governing equation and initial condition, then the upper bound of the normalized pressure is $y = 1$, and the lower bound is when the derivative vanishes, or $y = 0$. Similarly, the initial slope is $dy/dx = -1$ as one might anticipate within a normalized domain. From this behavior, one might select an exponential approximating function or

$$\tilde{y} = e^{-kz}.$$

Inserting this into the differential equation creates a residual,

$$\epsilon = -ke^{-kz} + e^{-2kz}.$$

By choosing the limits for the integral on the orthogonal relation to be $z = 0$ and $z = \infty$, we dramatically simplify the relationship for evaluating the constant $k = 2/3$. Thus, the Galerkin solution is

$$\tilde{y} = e^{-2z/3}. \tag{8.52}$$

8.7. Torque and Motion of the Finger

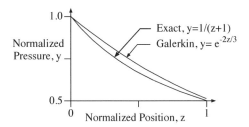

Figure 8.21. Comparison between exact and Galerkin solutions.

Figure 8.21 is a comparison between the exact solution, Eq. (8.50), and the Galerkin solution, Eq. (8.52). Note that the character of the solution is captured by the Galerkin solution, and the errors are not large. The normalized pressure at the end of the arterioles also is closely approximated.

It is left as an exercise to the reader to solve for the constant k from the orthogonal relation when the upper limit of the integral is chosen as $z = 1$ instead of $z = \infty$. With this choice of domain for optimizing the Galerkin approximating function, one expects smaller average errors over the problem domain.

8.6.3 Comments

This model has produced several important results. The original goal was to create a model that had the same basic autoregulatory character as demonstrated in nature. This goal was successfully achieved. The use of normalization, however, also provided a means of estimating the wall hoop stress, σ_w, that the smooth muscle-lined arterioles would have to regulate about to achieve this result. It is also interesting that the pressure exiting the arterioles will be one-half of the arterial pressure. This agrees well with experimental measurements in the capillary bed of functioning kidneys. A more detailed model for autoregulation of flow by the kidney is developed by Thompson [50] and the results compared to time-dependent experimental data.

8.7 TORQUE AND MOTION OF THE FINGER

The mathematical description of the mechanics of a finger can only be defined relative to the task being performed. In general, the vast majority of the motions of the human hand are quasistatic, and thus the inertial terms may be neglected. For concert pianists, however, inertial effects cannot be neglected. To quantify the behavior of the hand, therapists have studied the static torque-angle characteristics of the hand [10], and engineers have studied the dynamic character as shown in Fig. 8.22. As noted in Fig. 8.23, the torque-angle behavior of the finger is both nonlinear and speed dependent. The objective of this model is to better understand the dynamic behavior of the motion of the finger. We start by assuming a simplistic viscous model of the joint torque due to viscous resistance, T_c, of the form

$$T_c = C \frac{d\theta}{dt}$$

and a three-piece elastic model for the elastic stiffness, T_E, of the joint

$$T_E = T_0 + k(\theta)\theta.$$

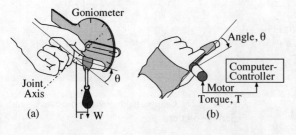

Figure 8.22. Pictorial of (a) static and (b) dynamic torque-angle testing of a finger.

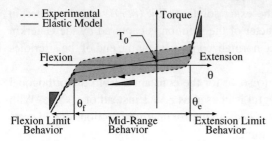

Figure 8.23. Torque-angle behavior of a finger joint.

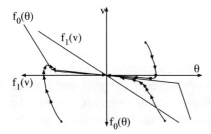

Figure 8.24. Transient response of a thumped finger.

These assumptions were experimentally demonstrated to be reasonably accurate by Llorens [27]. The resulting differential equation of the motion of the finger with no external torque and an initial displacement angle and velocity can be shown to be of the form

$$v\frac{dv}{d\theta} + f_1(v) + f_0(\theta) = 0, \qquad v(\theta_0) = v_0. \tag{8.53}$$

This initial condition may be thought of as thumping the finger. If the reader holds his left hand in a relaxed position (thus defining $\theta = 0$), and thumps it with a finger of the right hand, the resulting motion of the thumped finger is described by Eq. (8.53). The solution for this motion may be found using Pell's graphical method, and is depicted in Fig. 8.24. Note that the limiting behavior is governed by the elastic function, f_0. The terminal solution for every initial condition is $v = 0$, $\theta = 0$, indicating a very stable system. The solution is shown as a connected series of circular arcs resulting from the graphical approach used here. This model, although simplistic, compares favorably with the actual response of a thumped finger.

8.8 SUMMARY COMMENTS

The material presented in this chapter is not intended to cover all of the methods presented or all of the many types of systems covered in this text. What is presented is a variety of modeling and analysis problems that the reader can study for clues on organization, and exercises with a solution by the author. The reader should attempt to construct more detailed models, as well as other solution methods. This should be done independently from the material presented in this chapter if one is to master the concepts of analytical modeling. This also should help the student learn one of the principal lessons of this text: Never apply just one solution method. Only through comparison of results from various methods, tempered by a full understanding of the applicability of each method, can one gain the confidence needed to solve nonlinear problems.

If the reader has followed the material in this text, worked through the example and exercise problems in each chapter, then he has mastered the tools of modeling. What remains to be developed, and will be only be gained through the development and solution of many models, is a base of experience. The understanding of complex systems and the ability to become a master designer is enhanced by the lessons described in this book. Although there are many more solution methods, and much more to learn, the material encompassed within this text is the basis of modeling and analysis, both cornerstones of design.

8.9 PROBLEMS

The following problems are comprehensive modeling exercises that are intended to test the student's modeling as well as analysis skills. If symbolic manipulation programs such as MAPLE, REDUCE, or MATHEMATICA are available, each of the problems should take between 2 and 8 hours to complete, depending on the difficulty level of the problem.

8.1 *Pencil design.* Consider the design of a pencil for use by dressmakers. Rather than do a complete design,
   (1) Lay out all of the steps you expect to undertake in this process.
   (2) Develop a simple mathematical model of the process from the vantage of a corporate accountant interested in profit and loss. Assume that the model is to incorporate the entire sequence from the beginning idea through production and wholesale sales. Neglect distribution and retail sales.

   Keep your entire analysis and model development within a 2-hour period.

8.2 Create a mathematical model of the Earth for the purpose of predicting the rate of cooling of the molten core (dT_c/dt). A simplified physical model of the energy exchange of the Earth is shown in Fig. 8.25. In this model, the heat transfer rate through the crust may be approximated as a one-dimensional heat flow problem, or

$$\dot{q} = -kA\frac{\delta T}{h}.$$

Similarly, the radiation heat transfer from the Sun to the Earth is given as

$$\dot{q} = \sigma A \mathcal{F}_{es}\left(T_s^4 - T_{es}^4\right),$$

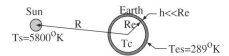

Figure 8.25. A simplified physical model of the Earth.

TABLE 8.5. Approximate Data for the Earth and Sun

Parameter	Value	Units
Radius of the Earth, R_e	2×10^7	ft
Specific heat of Earth's core, C_p	0.4	lbm/ft^3
Thermal conductivity of Earth, k	0.06	Btu/hr-ft-°F
Surface temperature of Sun, T_s	5,800	deg-K
Sun-Earth distance, R	93×10^6	miles
Avg. Normal Irradiance, I_0	1,400	w/m^2
Solar Angle, α_s	32	min
Geometry Factor, $\mathcal{F}_{es}$	0.000106	
Stefan-Boltzmann Constant, σ	0.1713×10^{-8}	Btu/hr-ft^2 deg-K^4
Effective temperature of space, T_{sp}	3.5	deg-K
Initial Earth surface temp, T_{es}	289	deg-K
Initial Earth core temp, T_c	1100	deg-K
Earth mantle thickness, h	5.28×10^6	ft
Earth surface area, A		ft^2
Mass of the Earth, M_e	12.2×10^{24}	lbm
Time, t		s

and a similar relationship holds for radiation heat transfer from the Earth's surface at T_{es} to space at an effective temperature of T_{sp}.

(1) Try to keep your model development and analysis time to 8 hours or less. Be certain to state your assumptions, expert knowledge, and restrictions in your model development.

(2) Sketch the solution for this problem.

(3) Estimate the drop in the Earth's temperature over the next 100 years. Use the data given in Table 8.5. Be forewarned: some of the data is missing and the units are not all consistent.

8.3 *Muscle motor design.* Design a muscle motor for a prosthetic system. The muscle is based on a simple mechanical principle of converting a rotational motion, θ, to a linear motion, x, as shown in Fig. 8.26. The muscle is to be directly driven by a Thompson Model A100 electric torque motor. The specifications for this motor are given in Table 8.6.

(1) Devise a math model of the mechanical muscle. Analyze the available force and excursion, given the specifications of the A100 torque motor. For this to be a useable muscle, it must be capable of producing a tension of 1000 lbf and an

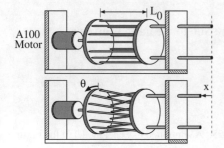

Figure 8.26. Pictorial view of a mechanical muscle. This concept is to be used to power a prosthetics device. The device provides a mechanical advantage for a simple torque motor and produces forces and excursions typical of human arm muscles.

8.9. Problems

TABLE 8.6. Critical Dimensions for the Thompson Academic Motors, Inc. Model A100 Actuator

Drawing item	Part description	Dimension (in.)
A	Motor diameter	1.375
B	Shaft diameter	0.249
C	Flange bolt separation	2.250
D	Flange length	3.250
E	Flange width	2.375
F	Flange bolt diameter	0.250

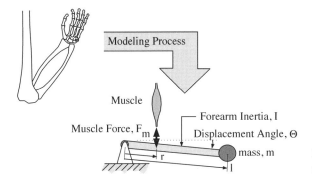

Figure 8.27. Schematic of a load on a forearm.

excursion of 1 in. Submit your model and plots of the maximal force-excursion capability of your design by electronic mail to your instructor.

(2) Plot the available torque versus the excursion of the muscle.

8.4 *Forearm muscle model.* Consider the effect of a sudden loading of a forearm by mass, m, as shown in Fig. 8.27. Develop a model for the forearm, assuming that the muscle force, F_m, is given as

$$F_m = F_0 \cos\left(\frac{2\pi r \theta}{l}\right)$$

and the damping torque, T_d, for the joint is given as

$$T_d = C_d(A + F_m \sin\theta)\dot{\theta}.$$

Develop and normalize the governing differential equation for this system, and describe all of the normalization constants. Be certain to state your transformations, assumptions, and show all clearly and succinctly. Demonstrate the model's performance. Use your own forearm as a source of data, and discuss your results.

8.5 *Airplane yaw-response model.* Consider the view of a simplified aircraft shown in Fig. 8.28. The yaw angle, β, of the aircraft is to be controlled by the rudder angle, α, which generates a lift, L, at the center of pressure. This is located on the rudder at a distance b from the rudder's pivot axis. The other fuselage dimensions (a, c and height w) are given in Table 8.7. The lift of the rudder is known to be a function of the rudder angle and the yaw angle of the aircraft,

$$L = C_L \tfrac{1}{2}\rho U_0^2 (\alpha - \beta),$$

TABLE 8.7. Parameter Values for the Aircraft Yaw Problem

Parameter	Value	Units
Air Properties		
Ref. specific Heat, C_{p0}	1.10	Btu/lbm-deg F
Ambient temperature, T_0	70.0	deg-F
Density, ρ	0.0785	lbm/ft^3
Velocity, U_0	500	ft/sec
Airframe Properties		
Rearward CG Distance, a	24	ft
Center of pressure, b	2	ft
Frontal CG distance, c	24	ft
Average section height, w	6	ft
Moment of inertia, I		ft-lbf-sec^2
Lift coefficient, C_L	1.25	
Drag coefficient, C_D	0.10	
Aircraft weight, W	1950	lbf

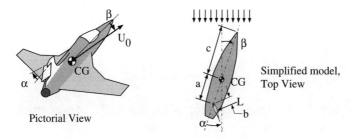

Figure 8.28. Simplified diagram of an aircraft in yaw.

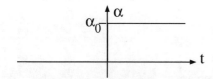

Figure 8.29. Forcing function for the yaw of an aircraft.

and the yaw damping moment is given as

$$T_D = C_D \frac{(a+b+c)^2 w}{3} \frac{1}{2} \rho \dot{\beta} |\dot{\beta}|.$$

The governing equation for the yaw angle of the aircraft then can be shown to be

$$C_0 \ddot{\beta} + C_1 \dot{\beta} |\dot{\beta}| + \beta = \alpha,$$

where $\beta(0) = \dot{\beta}(0) = 0$. To simulate a sudden change in the rudder angle, the rudder angle, α, is assumed to experience a step function of magnitude α_0 as shown in Fig. 8.29.

(1) Determine the analytical expressions for the coefficients C_0, C_1. Be careful to show your work.

8.9. Problems

(2) Transform the governing differential equation using the relations

$$y = \frac{\beta}{\beta_{max}},$$
$$x = \frac{t}{\tau},$$
$$F = \frac{\alpha}{\alpha_{max}}$$

to obtain the following:

$$\frac{d^2y}{dt^2} + \left|\frac{dy}{dx}\right|\frac{dy}{dx} + y = F.$$

Be certain to state clearly your choices for the scaling values $\alpha_{max}, \beta_{max}, \tau$.

(3) Obtain a numerical solution for the yaw motion of the aircraft subject to the given initial conditions and to a step change in α.

(4) Obtain a solution for the yaw motion as above, but using a graphical technique.

Show your development concisely and in a manner that attests to your knowledge of the subject matter. Produce the graphics on a workstation using the graphical support library.

8.6 *Automobile engine model.* Construct a mathematical model of an automobile engine, as shown in Fig. 8.5. Model the fuel-and-air delivery system. Assume the input variable to be the throttle angle, α, and engine speed, N. Assume that the fuel/air ratio, $\mathcal{FA}$, is a constant. The pressures at varying throttle angles and engine speeds are given in Table 8.1 where $N_{idle} = 750$ rpm and $N_{max} = 6000$ rpm. The engine is a rotating inertial system with a variable load torque, T_L, and inertia, I_0. The combustion process converts the inlet fuel/air mixture to useful power, but much of that energy leaves the exhaust system as waste heat. Assume that the fuel energy is used to heat the cylinder at constant volume and other processes normally included in an "Air Standard Analysis." The dynamic (not static) bearing torque and the piston-to-cylinder frictional torque are to be modeled as acting together to produce a loss torque proportional to the engine speed, N. (Note, this may be symbolically stated as $C_0 + C_1 N$.)

Each component of your model (carburetor, frictional losses, etc.) is to be represented using separate subroutines which together represent the solution for the response of the engine. Scale the model such that it represents a six-cylinder engine capable of producing a net power of 140 hp at 6,000 rpm. Also note that the engine will idle at 750 rpm when the throttle angle is at $\alpha = 0°$ and there is no load on the engine.

State all of the assumptions you make in your derivations and produce a report showing the resulting model and the governing equations. Produce your graphics using a graphics program or library.

Demonstrate the model's performance in the following manner:

(1) Derive an expression for and plot the pressure torque versus crankshaft angle, θ, during the piston travel from ignition (TDC) to the opening of the exhaust valve at the bottom of the piston travel (BDC). Integrate this function to determine the

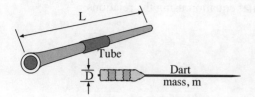

Figure 8.30. Schematic of an early blowgun.

(average) mean effective combustion pressure, P_c, over the expansion process following ignition. Plot P_c versus compression ratio over the range $r = 7$ to 15.

(2) Assume that the engine is loaded to steady state at varying rotational speeds. Note that these maxima are steady-state values and this greatly simplifies the solution of the differential equations describing the system. Graphically depict the full-throttle maximum output horsepower, P_L, and torque, T_L, versus engine rotational speed, $N = 0$ to 8000 rpm, graphically.

(3) Solve for the engine speed versus time if there is no load on the engine ($T_L = 0$) and the throttle is snapped from idle setting to full throttle and held there. Depict this result graphically over the range $N(\alpha_0, 8000)$ rpm.

Use the data and variable symbols given in Table 8.3 for your derivations and calculations:

8.7 *Blowgun model.* In the archives of the Maxwell Museum of the University of New Mexico, there are several specimens of early blowguns. These long tubular weapons are shown schematically in Fig. 8.30. The dart is a long sharpened stick with a soft rear seal similar to cotton. This is inserted into the blowgun tip first, and a quick, hard puff of air accelerates the dart down the tube and out toward some target. Develop a model of the motion, v, of the dart in the blowgun tube as outlined below. In your calculations, take extra care with your units.

(1) Derive the one-dimensional governing differential equation for the velocity, $v(t)$, of the dart within the blowgun tube. Assume that the friction between the dart and the tube is given by Drag $= Cv^3$, where the frictional coefficient $C = 0.2618 \times 10^{-5}$ lbf-s^3/ft^3. Also, assume that the following data are in effect: The pressure that the user can develop is $P = 2$ lbf/in.^{2}, the tube diameter is $D = 0.6$ in., and the dart mass is $m = 0.2$ lbm. Please note your assumptions.

(2) Normalize the differential equation. From this effort, find a formula for the maximum velocity, v_{max}, of the dart. Numerically solve for this velocity.

(3) How would you choose the optimal length, L, of the blowgun tube? Explain the rationale behind your approach.

(4) Solve the full nonlinear problem in as many ways as you can, but using at least one of the approximate methods.

8.8 *Canoe model.* Create a mathematical model of a canoe as shown in Fig. 8.31. The objective of this exercise is to discern the maximum speed of the canoe and how long it takes to reach this velocity when starting from rest. Be certain to state your assumptions and cite references for your work.

8.9 *Class model.* Attempt to mathematically model this course. Don't attempt to find any solutions to your model; just work on detailing the model features. Show your development concisely and clearly.

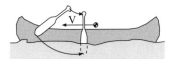

Figure 8.31. Simple canoe.

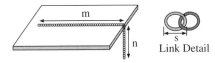

Figure 8.32. Diagram of a chained system of rigid elements under the influence of gravity and friction.

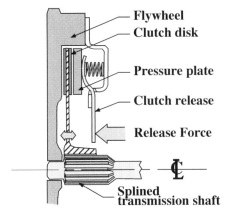

Figure 8.33. Expanded view of an automotive clutch assembly.

8.10 *Falling-chain model.* In the system shown in Fig. 8.32, m elements of a chain of rigid links are initially at rest on the table and n elements are hanging below the edge of the table. Each element has a weight w_i and length s. Assuming that the friction between each element and the table surface is proportional to its normal force and its velocity, derive the governing equations for the motion for the system as a continuum and as a set of discrete elements. Be certain to note your assumptions.

8.11 *Automotive clutch performance.* The analysis of the performance of an automotive clutch is dependent on many variables. These include but are not limited to
- the torque that it is required to transmit,
- the maximum force and travel of the driver's foot pedal,
- the maximum rotational speed that the clutch will be exprected to see during its service life,
- the desired inertia of the clutch to contribute to the smoothness of the engine.

A typical automotive clutch assembly is shown in Fig. 8.33. Note that the flywheel is to bolt directly onto the engine's crankshaft at its left end, and the transmission input shaft is on the right-hand side. Model the dry clutch assembly and derive the relationship between its input and output torque and power. Model the frictional surface and materials to withstand the inertial and torque loads that the clutch will sustain. List all of your assumptions at the beginning of your model development.

A component of your model must include the thermal effects associated with slipping of the clutch during startup of an automobile from rest. In this model, assume that the frictional coefficients of the system remain constant up to a critical temperature, at which time the frictional coefficients are reduced by one order of

TABLE 8.8. Clutch Property Values and Dimensions

Description	Value	Units
Clutch facing, asbestos		
Specific heat, C_{pa}	0.20	Btu/lbm-°F
Density, ρ_a	36	lbm/ft^3
Flywheel, steel		
Specific heat, C_{pa}	0.11	Btu/lbm-°F
Density, ρ_a	490	lbm/ft^3
Thickness, h	1.0	in.

TABLE 8.9. LSU Dragster Performance Data

t (s)	x (ft)	v (mph)
0.00	0	0
1.30	60	
3.60	330	
5.70	660	
8.90	1320	146

Figure 8.34. Econorail dragster.

magnitude because of melting and glazing of the clutch's frictional surface. This temperature is approximately 1200°F for some materials.

As a test of the model, assume that the input shaft is initially rotating at 6000 rpm, with the output shaft at rest but with the clutch disengaged. At time $t = 0$, the clutch is engaged and the output shaft torque results from an inertial load of 485 ft-lbf-s^2. Assume that the motor is capable of delivering a constant torque of 150 ft-lbf. For the specifications given in Table 8.8, plot the rotational speed of the output shaft and the temperature of the clutch versus time. Select a clutch diameter and other clutch properties to ensure that the clutch does not reach an operating temperature of 1000°F.

8.12 *Rail dragster model.* In 1992, a senior design team at Louisiana State University designed, constructed, and tested an econorail gas dragster. Create a mathematical model of this dragster as shown in Fig. 8.34, and minimize the run down the $\frac{1}{4}$-mile track. The vehicle is to start from rest and, to calibrate your model, the actual performance of LSU's dragster is given in Table 8.9. The transmission used was a Powerglide two-speed with a maximum gear ratio of 10.5:1 powering the 23-in.-diam rear tires. The engine was a Chevrolet LS-6, Mark 4 big block. It produced its maximum torque of 480 ft-lbf at 4000 rpm, and its maximum power of 450 hp at 5300 rpm. The gear ratio of the differential was 4.9:1. The frontal area of the 1750-lbm vehicle was measured from a photograph at 3.34 ft^2, and the rolling

8.9. Problems

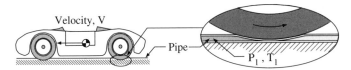

Figure 8.35. Energy production system for off-ramps.

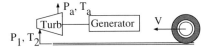

Figure 8.36. Flow diagram of proposed energy retrieval system.

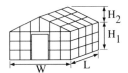

Figure 8.37. Glass house for use in home gardening.

resistance was found to be directly proportional to the vehicle weight:

$$R_{roll} = 0.015W.$$

The aerodynamic drag coefficient was approximately $C_D = 0.6$. Be certain to state your assumptions and cite references for your work.

8.13 *Energy recovery system model.* Consider a proposed energy retrieval system as shown in Fig. 8.35. Derive the differential equation for the system, portrayed schematically in Fig. 8.36, assuming that the turbine work is given by

$$W_{1-a} = C_p(T_1 - T_a).$$

Derive the one-dimensional governing differential equation for the temperature in the collapsable pipe, T_1. Assume that the pipe is filled with air that behaves as an ideal gas. Note any other pertinent assumptions. Be certain to show your work clearly, and in a manner that attests to your knowledge of the subject matter.

8.14 *Glass house model.* Model a glass house (greenhouse) to be used by individuals in the winter for storing and raising plants. The purpose of the model is to serve as a controller of a bang-bang heater during the winter. The physical layout of the greenhouse is shown in Fig. 8.37. Be certain to write down all of your assumptions. The greenhouse is to be built from glass having a thermal conductivity given by $k = 0.0081[1+0.02T]$, where the temperature, T, is in absolute K and the resulting thermal conductivity, k, is in W/m-K.

(1) Construct a mathematical model for the temperature in the greenhouse, considering thermal conductivity alone (no radiant energy exchange). Be certain to normalize you model equations.

(2) Compute the minimum power necessary to keep the greenhouse at a temperature no less than 289 K.

(3) Assuming that the heater has only two states, off and on (bang-bang), plot the internal air temperature as a function of time for mean outside air temperatures of 250 K, 270 K, and 290 K with a daily variation amplitude of ±5 K. The air

temperature, T_{air}, varies with time (t in days) as given by

$$T_{air} T_{mean} + T_{var} \sin\left[\frac{\pi(t-360)}{770}\right].$$

(4) Determine a characteristic time constant for the greenhouse if such a thing exists.

(5) Repeat that above model and computations including solar radiant heating effects given by the equation below for an absorptivity, $\alpha = 0.4$:

$$E_{in} = 20\alpha \sin\left[\frac{\pi(t-360)}{770}\right].$$

The incident solar energy flux, E_{in}, is in W/m²-hr based on the planform area of the greenhouse.

8.15 *Lightbulb model.* Create a mathematical model of a lightbulb and predict the time change of the surface temperature of the glass bulb. Be certain to state all of your assumptions and the modeling equations you used to derive the governing equations. Remember to start *simple* and only grow as complex a model as is absolutely necessary. Try and keep your total model development and analysis time to 8 hours or less.

After you have developed a formula to predict the surface temperature, estimate dimensions and other physical parameters for a 60-W light and plot your results. You may choose a symbolic algebra tool to aid in obtaining a solution, or write your own analysis code.

8.16 Learn a symbolic algebra language such as MAPLE, REDUCE, or MATHEMATICA and use it to solve the following problem. You are given the following equation for a three-dimensional line in space:

$$\frac{x-x_1}{x_2-x_1} = \frac{y-y_1}{y_2-y_1} = \frac{z-z_1}{z_2-z_1};$$

and similarly, an equation for a plane:

$$x + y + \frac{z}{2} = c.$$

(1) Find the point of intersection, $P(x, y, z)$, of the line and the plane. Generate a file portraying your solutions directly out of the symbolic algebra engine.

(2) What limitations are there on your analytic solution?

(3) What are the numerical values of the intersection coordinates given $P_1 \to (x_1 = y_1 = z_1 = 0)$, $P_2 \to (x_2 = 1, y_2 = 2, z_2 = 1)$ and $c = 25$? Again, use the symbolic algebra system to directly write out your results.

8.17 *Helicopter model.* Construct a mathematical model of a helicopter as shown in simplified form in Fig. 8.38. Assume that the lift and drag of both the main rotor

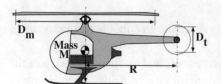

Figure 8.38. Simplified model of a helicopter.

disc and the tail rotor disc are linear with the angle of attack of the blades, α_m and α_t, respectively. The standard aerodynamic formulas for lift and drag are

$$L = \tfrac{1}{2}\rho V^2 C_L A \alpha,$$
$$D = \tfrac{1}{2}\rho V^2 C_D A \alpha,$$

where A is the area of the disc based on its mean effective diameter, D_{eff}, ρ is the mass density of air, and V is the net velocity of the blades relative to the surrounding air. This can be approximated as

$$V \approx \sqrt{\left[\frac{N D_{\text{eff}}}{2}\right]^2 + \left[\frac{dy}{dt}\right]^2},$$

where y is the vertical position of the helicopter, dy/dt is the vertical velocity, and the disc rotational speed is N. The $\dot{y}$-velocity correction also influences the angle of attack, α, according to

$$\alpha_{\text{eff}} = \alpha - \tan^{-1}\frac{\dot{y}}{V}.$$

Your model should attempt to predict the motion and position of the helicopter vertically and rotationally for given operating rotor speeds and blade angles. State all of the assumptions you make in your derivations and produce a report showing the resulting model and the governing equations. Use a minimum of three different solution methods to validate your results. One of these methods must be your own Runge-Kutta numerical integration scheme. Demonstrate the model's performance in the following manner:

(1) Determine the design condition necessary just to keep the helicopter rotationally stable. Assume that the ratio of tail rotor speed to main rotor speed is held constant using a gearbox. The main- and tail-rotor angles of attack are to be kept roughly equal, but varied slightly to maintain rotational control. This equates to $\alpha_t = \alpha_m + \delta$, where δ is the control variable. Compute the gearbox ratio and find an equation for the tail-rotor control variable, δ, in terms of the system variables.

(2) If the helicopter is held in a hovering mode with the main rotor at $N_m = 400$ rpm and $\alpha_m = 5°$, compute the helicopter's weight.

(3) Plot the torque and horsepower required by the main and tail rotors over the range of input shaft speeds $N_m = 0$ to 800 rpm, and angles of attack over the range $\alpha = 0$ to $30°$. Assume equilibrium conditions ($\omega = \dot{\omega} = \dot{y} = 0$).

(4) If the helicopter blade angles are suddenly raised to $\alpha = 10°$ and the main rotor is at $N_m = 400$ rpm while the helicopter is sitting at rest on the ground, compute its resulting vertical motion.

Use the data and variable symbols in Table 8.10 for your derivations and calculations.

8.18 *Ping-pong model.* Create a mathematical model to predict the height, H, and distance, L, traveled by a ping-pong ball after it is launched by the device shown in Fig. 8.39. Scale your model using the numerical data given in Table 8.11. The actual

TABLE 8.10. Helicopter Property Data[a]

Variable	Value	Units
Air density, ρ	0.075	lbm/ft^3
Helicopter mass, M		lbm
Helicopter inertia, I_0	45,000	ft-lbf-s^2
Engine power rating, P		hP
Main-to-tail rotor separation, R	18.0	ft
Main rotor lift coefficient, C_{Lm}	1.0	
Main rotor drag coefficient, C_{Dm}	0.01	
Tail rotor lift coefficient, C_{Lt}	1.5	
Tail rotor drag coefficient, C_{Dt}	0.01	
Main rotor effective diameter, D_m	22.0	ft
Tail rotor effective diameter, D_t	4.0	ft
Tail rotor control variable, δ		deg
Gravitational constant, g_c	32.2	ft-lbm/lbf-s^2

[a] The data in this table are tentative and you may change them to make the problem more realistic. If you use an existing helicopter as your model, state where you obtained your data.

TABLE 8.11. Numerical Data and Symbol Definition for the Ping-Pong Launcher

Variable	Value	Units
Launcher stroke, S	9.0	in.
Launch angle, θ	45.0	deg
Spring free length, L_0	11.5	in.
Spring weight, M_s	0.18	lbm
Spring stiffness, k	2.5	lbf/in.
Ping-pong ball mass, m	2.6	oz.
Ball diameter, D	1.5625	in.
Horizontal travel, L		in.
Vertical height, H		in.

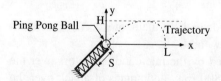

Figure 8.39. Ping-pong ball launcher.

measured horizontal travel of the ping-pong ball was 18.5 ft. Discuss any differences between your result and this experimental result.

8.19 *Sail model.* Create a mathematical model to predict the time required to raise the sails on the America Cup racing yacht *Thompsonia*, as shown in Fig. 8.40. It is of great interest to select the proper gear ratio for the sail mechanism shown. The sail is triangular and weighs γ lb/ft^2. When it is fully raised, it is W wide and H high. One important aspect of your model is how you have chosen to represent human performance.

8.9. Problems

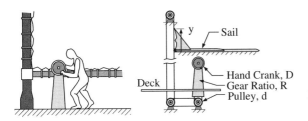

Figure 8.40. Raising the sail on the *Thompsonia*.

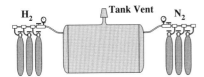

Figure 8.41. Tank dilution problem.

8.20 *Tank dilution model.* Consider the tank, shown in Fig. 8.41, initially charged with 100% nitrogen at standard temperature and pressure. If a connected hydrogen source is allowed to bleed hydrogen into the tank with the vent opened, how much hydrogen will be required to reduce the concentration of nitrogen to 50% by volume?
  (1) Develop a mathematical model for the above problem. Be certain to state the assumptions you made in developing this model. This is to be a mathematical (abstract) model; no numbers here, please.
  (2) State whether your model will predict an upper limit to the H_2 added or a lower limit, or in between.
  (3) Recommend enhancements to the model you have given that will improve its accuracy.
  (4) Use your model to predict the numeric value of the H_2 added. Assume the tank volume to be 20 ft^3 at initial conditions of 1 atm, 530°R.

8.21 *RC circuit model.* Consider the mathematical model for the RC circuit given in Chapter 2, Example 2.7.
  (1) Derive the governing differential equations for the temperature change and the current through the resistance. Neglect all current to the output, E_0.
  (2) Normalize the governing differential equations using the following transformations:
  $$\theta = \frac{T}{T_0}, \qquad I = \frac{i}{i_0}, \qquad y = \frac{t}{\tau}.$$
  (3) Obtain the differential equations for dI/dy and $d\theta/dy$ as driven by the forcing function, E_I, or its derivative.
  (4) Take the solution of this system as far as possible. Use a minimum of two solution techniques and produce a report of your findings.
  Use the data given in Table 2.1 for your calculations. Produce the graphics on a workstation using a standard graphics program.

8.22 *Underwater-launched rocket model.* Model the two-dimensional dynamics of a rocket launched from an underwater platform, as shown in Fig. 8.42. You are expected to make estimates of many of the rocket's properties including its weight,

TABLE 8.12. Properties of Submerged Launch Rocket

Variable	Value	Units
Length, L	30	ft
Diameter, D	4	ft
Init. velocity, V_0	0	ft/s^2
Range, S	150	miles
Initial weight, W		lb
Thrust, T		lb
Nozzle exit velocity, V_e	1500	ft/s
Empty vehicle weight, w	5000	lb

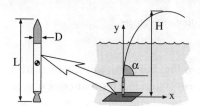

Figure 8.42. Launching of an underwater rocket.

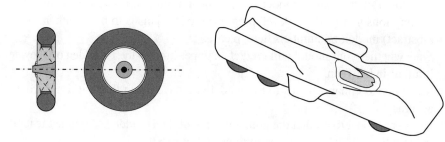

Figure 8.43. The Mercedes Benz T-80 Land Speed Record Vehicle. The T-80 was designed by Dr. Ferdinand Porsche in 1937, completed in late 1939, but never tested or run at speed because of World War II.

drag coefficients, and shape. Table 8.12 provides some scaling information for your assistance.

(1) Develop a mathematical model to predict the maximum height, H, and trajectory, $[x(t), y(t)]$, for various launch angles, $\alpha = [0, 30°]$. Include varying drag with elevation due to changing air density, ρ_a, and variation of the rocket vehicle weight as its fuel is consumed. Be certain to write down all of your assumptions.

(2) Normalize the model. You may need to change the normalization values over each phase of the flight (e.g., underwater, air).

(3) Provide graphical plots of the trajectory for varying launch angles and fuel load.

8.23 *LSR tire model.* Create a mathematical model for the growth of a high-speed tire with speed and pressure. The tire system you are to model is the 32×7 tire for the T-80 Porsche and Mercedes Benz Land Speed Record car, depicted in Fig. 8.43. The tire was rated at 350 mph, but its dimensions varied significantly with speed because of the inadequate tire technology of 1938. The specifications for this special

8.9. Problems

TABLE 8.13. Experimental Data on Porsche/Mercedes T-80 LSR Tire[a]

Ground speed (mph)	Section width (in.)	Tread width (in.)	Outer diameter (in.)
0	8.268	4.10	47.09
249	7.126	4.06	48.39
348	6.299	3.79	48.86

[a] Data extrapolated from circa 1938 experimental work. The tire section has a thickness at the tread estimated at 10 mm (0.394 in.). The diameter shown above is the maximal diameter (mid-tread) of the tire. All measurements listed are at the design inflation pressure of 80 psig.

tire are included in an article by Batchelor [5]. Some pertinent information on the tire's construction and experimental test has been estimated from data in this article and is given in Table 8.13. Based on this information, perform the following tasks:

(1) Plot the experimental data presented by Batchelor [5] for tire growth with speed and postulate a governing relationship for this effect in terms of the internal gage pressure, strain, and properties of the tire.

(2) Derive the governing time-dependent differential equations for the tire, and predict the steady-state response of the tire to speed changes.

(3) Solve the differential equation for the dynamic response of the tire to an incremental wheel velocity change from 250 to 280 mph resulting from a sudden loss of traction at speed.

Produce a report, complete with a discussion about how you would compute the dynamic response of the tire to speed changes and any problems you foresee in the model. In your derivations, be certain to state all of the assumptions you make. Some of these you should include are the following:

- Neglect the effects of all internal tire stresses on the radius except those involving the circumferential stress.
- Assume that the tread and carcass have uniform density throughout the tire with an average value of 75 lbm/ft^3.
- Assume Poisson's ratio to be zero (explain the effect of this assumption on the tire thickness, h, and tread width, w).
- Assume that the stress-strain relationship is exponential about the design *static* inflated condition (80 psig).

8.24 *Transmission model.* Construct a mathematical model of an automobile transmission, shown pictorially in Fig. 8.44. Note that gears B and C slide together to engage and disengage respectfully.

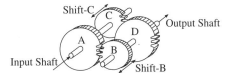

Figure 8.44. An automotive transmission showing the various gear ratios.

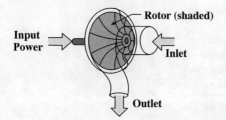

Figure 8.45. Typical automotive water pump.

8.25 *Automotive water pump model.* Model an automobile water pump and produce a routine that will take engine speed as an input and compute several variables of interest. For your use in modeling, a typical view of a water pump is shown in Fig. 8.45. The subroutine shall have the following input variables:

N = engine speed, rpm;
Q = water flow rate, gpm;
T = shaft torque, ft-lbf;
W = power required, Hp;
h = pressure head, ft.

Be certain to specify any parameters of the pump design and provide approximate values for them. Include with your routine a main program to call it so it may be tested. You are encouraged to look up information in the library. You will find that papers from the Society of Automotive Engineers (SAE) might prove especially useful. If you use any work done by others, you must include full reference to their work.

Use the following approximate data to size your example pump:

$Q \approx 2{,}000$ gal./min,
$W \approx 5$ Hp,
Press head, $h \approx 10$ ft.

8.26 *Swimming whale model.* Model the one-dimensional dynamics and the energy balance of the whale shown in Fig. 2.9. Be certain to write down all of your assumptions. The whale is to be considered as a system. Its tail, when used as a propulsive motor, generates a thrust, T, along its major axis which is given as

$$T = T_0 |\sin(\omega t)|.$$

There is an associated drag force on the whale as it moves at velocity, v, through the seas, which have a fluid density, ρ, which can be written as

$$D = C_d A_F \tfrac{1}{2} \rho v^2,$$

where the maximum cross-sectional area is A_F. The whale gathers food while swimming at a rate proportional to its velocity. Thus, the energy rate of increase of the whale is

$$\dot{E}_{\text{in}} = \eta_{\text{th}} F A_F v,$$

where γ is the energy (nutritional) content of the food the whale consumes. Assume the whale is of neutral buoyancy and has an approximate mass of $m = \tfrac{2}{3} \rho A_F L$. Further assume that the whale's energy is conserved and heat transfer can be neglected. Then, the one-dimensional governing equations of motion and of energy are

$$\rho C_p \frac{d\theta}{dt} = \dot{E}_{\text{in}} - \dot{W}_{\text{tail}} - Dv$$
$$= \eta_{\text{th}} F A_F v - T_0 |\sin \omega t| - C_D A_F \tfrac{1}{2} \rho v^3, \tag{8.54}$$

$$\frac{2}{3}\rho A_F L \frac{dv}{dt} = g(v, t). \tag{8.55}$$

(1) As a first approximation, compute the *average* value of the thrust, $\bar{T}$.
(2) Assuming that the thrust is a constant, determine the function $g(v, t)$.
(3) Find an analytical expression for the maximum attainable velocity. State the conditions that you feel define this condition.
(4) Determine an analytical expression for the maximum sustainable velocity. State the conditions that you feel define this condition.
(5) If possible, apply Picard's method to determine a first approximation to the velocity, v. Assume that the whale starts from rest, an initial condition $v(0) = 0$.
(6) Assume an approximation function, $\tilde{v} = C_0 + C_1 e^{-kt}$, and use Galerkin's method to solve the problem analytically.

8.27 *Radiation heat transfer analysis.* In Chapter 3, Section 3.2, a model of thermal radiation transfer was presented along with an exact solution. Normalize Eq. (3.53) using the transformations $\theta = T/T_i$, $x = t/\tau$ and find the time constant, τ, required to reduce the governing equation to

$$\frac{d\theta}{dx} = \theta^4 = 1, \qquad \theta(0) = \theta_0.$$

Next, assume a Galerkin solution function that satisfies the initial condition

$$\tilde{\theta} = \theta_0 + (1 - \theta_0)(1 - e^{-Cx})$$

and use the Galerkin method to find the analytic solution for the unknown constant, C. Then, compare this solution to the exact solution for the three cases, $\theta_0 = 0.1, 0.01, 0.001$.

8.28 *Historical figures.* Write a one-page summary of the life and accomplishments of each of the following historical figures in science and mathematics:
- Leonardo da Vinci;
- Euler, father of the Euler method of integration;
- Taylor, of Taylor-series fame;
- Picard, who originated Picard's method;
- Runge, one of developers of the Runge-Kutta numerical integration method;
- Kutta, the other one;
- Liènard, originator of the Liènard's method;
- Pell, of Pell's method fame;
- Galerkin, originator of Galerkin's method.

This is due as both a hard copy and an electronic file. Please electronically mail this file to your instructor.

8.29 *Cliff climbers.* Two climbers of equal mass are attached by an inelastic, massless rope of length, L (see Fig. 8.46). One slips off the edge of the cliff at time $t = 0$.

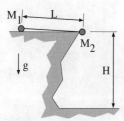

Figure 8.46. Simplified representation of falling climbers.

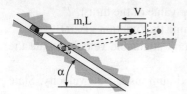

Figure 8.47. Rod undergoing constrained motion.

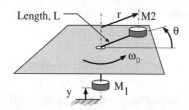

Figure 8.48. Physical arrangement of two masses connected by a massless wire.

The second cannot get a grip (no friction), and starts to slide as well. How high is the cliff if both climbers land at the same time?

8.30 *Rod undergoing constrained motion.* The right end of the rod shown in Fig. 8.47 is forced to move from right to left with a constant velocity, B. What is the bending moment at the center of the rod when it is horizontal?

8.31 *Masses connected by massless wire.* Consider the two masses connected by a thin massless wire of length L, shown in Fig. 8.48. The upper mass, M_1, is constrained to move horizontally on a frictional plate with a coefficient of friction of μ. The lower mass, M_2, is assumed to only move vertically. Initially, both masses are at rest, but the plate is spinning with a constant angular velocity, ω_0. If these masses are assumed to be point masses and the hole in the plate of negligible diameter, derive the governing equations of motion and obtain solutions. Describe any unusual behavior, limit cycles, or other characteristic scales or times.

8.32 *Pogo stick model.* Design an alcohol-powered Pogo Stick. Include a model of the thermodynamic processes, the physical device, and a simple model of the mechanics of the rider.

In your report, include a theoretical development and a solution (simulation) to support your design. This solution should enable you to investigate how the device will adapt to the *size* and *weight* of the rider and to variations in the *throttle setting*. Also include drawings with critical dimensions specified.

8.33 *Viscosimeter design.* Design a viscosimeter based on the principle of a falling sphere. The concept is that the terminal velocity of a sphere through a medium is a function of the viscosity of the medium. The equation of motion may be shown

8.9. Problems

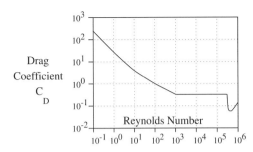

Figure 8.49. Variation of drag coefficient for a sphere with Reynolds number.

to be

$$m\frac{dV}{dt} = C_D A \left[\frac{1}{2}\rho V^2\right] - mg, \quad (8.56)$$

where some of the important variables are m = mass of the sphere; A = cross-sectional area of the sphere; y = vertical position of the sphere; V = vertical velocity, dy/dt, of the sphere; t = time of fall; g = local acceleration of gravity; C_D = drag coefficient; ρ = mass density; and μ = viscosity. The drag coefficient is not a constant, however, but dependent on the Reynolds number, which varies with the velocity. The Reynold's number is given as

$$Re = \frac{\rho D V}{\mu}. \quad (8.57)$$

Experimental data for the drag coefficient are shown in Fig. 8.49. Design an experimental device capable of measuring viscosity over a range of $10^{-6} \geq \mu \geq 5 \times 10^{-5}$. This must be a tabletop device with a maximum height of 2 m.

In your report, include the theoretical development and a solution (simulation) to support your design. Also include drawings with critical dimensions specified. To demonstrate the behavior of your model, please include plots of velocity versus distance for a falling steel sphere over a diameter range of 0.5 in. $\geq D \geq$ 2.0 in.

References

[1] Abbott, D. E., *Course Notes, Boundary Layer Theory,* Purdue University, W. Lafayette, IN, 1967.
[2] Abramowitz, M. and I. A. Stegun, *Handbook of Mathematical Functions,* U.S. Department of Commerce, Washington, DC, 1965.
[3] Baker, G. L. and J. P. Gollub, *Chaotic Dynamics, an Introduction,* Cambridge University Press, New York, 1990.
[4] Barbashin, E. A. and N. N. Krasovski, "On the Global Stability of Motion," *Proc. Academy of Science, USSR, Vol. 86,* 1952, pp. 453–6.
[5] Batchelor, D., "The Porsche Designed, Mercedes Benz T-80," *Porsche Panorama,* Porsche Club of America, Alexandria, VA, Oct. 1990.
[6] Bellman, R., *Perturbation Techniques in Mathematics, Physics, and Engineering,* Holt, Rinehart, and Winston, 1964.
[7] Bell, D. J., *Mathematics of Linear and Nonlinear Systems for Engineers and Applied Scientists,* Oxford University Press, New York, 1990.
[8] Bird, R. B., W. E. Stewart, and E. N. Lightfoot, *Transport Phenomena,* Wiley, New York, 1960, p. 73.
[9] Blaquiere, A., *Nonlinear System Analysis,* Academic Press, New York, 1966.
[10] Brand, P. W. and A. Hollister, *Clinical Mechanics of the Hand,* 2nd ed., Mosby Press, St. Louis, MO, 1993, Chapters 5 and 7.
[11] Cesari, L., *Asymptotic Behavior and Stability Problems in Ordinary Differential Equations,* 3rd ed., Academic Press, New York, 1971.
[12] Cheney, E. W. and D. R. Kincaid, *Numerical Mathematics and Computing,* Brooks/Cole, Pacific Grove, CA, 1994.
[13] Chen, S. H., Y. K. Cheung, and S. L. Lau, "On Perturbation Procedures for Limit Cycle Analysis," *Int. J. Non-Linear Mech., Vol. 26, No. 1,* 1991, pp. 125–33.
[14] Chestnut, H. and R. W. Mayer, *Servomechanism and Regulating Systems Design, Vol. 1,* Wiley, New York, 1959.
[15] Courant, R. (ed), *Studies in Nonlinear Vibration Theory,* New York University, Inst. for Mathematics and Mechanics, New York, 1946.
[16] Fiacco, A. V. and G. P. McCormick, *Nonlinear Programming: Sequential Unconstrained Minimization Techniques,* Classics in Applied Mathematics, Society for Industrial and Applied Mathematics, Philadelphia, 1990.
[17] Dixon, J., *Design Engineering: Inventiveness, Analysis, and Decision Making,* McGraw-Hill, New York, 1966.
[18] Galilei, G., *A Discourse Concerning the Natation of Bodies upon and Submersion in the Water,* Presented to Don Cosimo, Great Duke of Tuscany, Italy, 1611.
[19] Jahnke, E. and F. Emde, *Tables of Functions with Formulae and Curves,* 3rd ed., B. G. Teubner, 1938, reprinted by McGraw-Hill, New York, 1960.
[20] Johnson, W. C., *Mathematical and Physical Principles of Engineering Analysis,* McGraw-Hill, New York, 1944.
[21] Kantorovich, L. V. and V. I. Karlov (Translated from Russian by Curtis D. Benster), *Approximate Methods of Higher Analysis,* Interscience, 1964.

[22] King, J. T., *Introduction to Numerical Computing*, McGraw-Hill, New York, 1984.
[23] Koberg, D. and J. Bagnall, *The Universal Traveller: A Soft Systems Guide to Creativity*, Kaufmann, Los Altos, CA, 1974.
[24] Krylov, N. and N. Bogoliubov, *Introduction to Nonlinear Mechanics*, Princeton University Press, Princeton, NJ, 1943.
[25] Lefschetz, S., *Differential Equations: Geometric Theory*, 2nd ed., Wiley, New York, 1963, pp. 30–9.
[26] Legendre, A.-M., *Traité des Fonctions Elliptiques*, Paris, 1825; reprinted as *Tables of the Complete and Incomplete Elliptic Integrals*, Cambridge University Press, University College, London, 1934.
[27] Llorens, W. A., "An Experimental Analysis of Finger Joint Stiffness," M.S. thesis, Dept. of Mechanical Engineering, Louisiana State University, Baton Rouge, LA, May 1986.
[28] Lyapunov, A. M., "Obshchaya zadacha ob ustoychivosti dvizhenia" ("The General Problem of the Stability of Motion"), Ph.D. dissertation, Moscow University, 1982. Reprinted as No. 17 in the Annals of Mathematics series, Princeton University Press, Princeton, NJ, 1947.
[29] Moody, L. F., "Friction Factors for Pipe Flow," *Trans. ASME, Vol. 66, No. 8*, Nov. 1944, pp. 671–84.
[30] Mosely, F. A. and R. M. Evan-Iwanovsky, "The Effects of Non-Stationary Processes on Chaotic and Regular Responses of the Duffing Oscillator," *Int. J. Non-Linear Mech., Vol. 26, No. 1*, 1991, pp. 125–33.
[31] Nutting, W., *The Clock Book*, Garden City Publ., New York, 1976, p. 308.
[32] Perli, S., *Theory of Matrices*, Addison-Wesley, Cambridge, MA, 1952.
[33] Poincaré, H., *Les Méthodes Nouvelles de Mécanique Céleste*, 3 volumes, Gauhier-Villars et fils, Paris, 1893.
[34] van der Pol, B., "Forced Oscillation in a Circuit with Nonlinear Resonance," *Philos. Mag., Vol. 3*, 1927, pp. 65–80.
[35] Powell, M. J. D., "An Efficient Method for Finding the Minimum of Several Variables without Calculating Derivatives," *Computing J., Vol. 7, No. 2*, 1964, pp. 155–62.
[36] Press, W. H., B. P. Flannery, S. A. Teukolsky, and W. T. Vettering, *Numerical Recipes*, Cambridge University Press, Cambridge, UK, 1986.
[37] Rothe, C. F., F. Nash, and D. E. Thompson, "Patterns in Autoregulation of Renal Blood Flow in the Dog," *Am. J. Physiol., Vol. 220*, June 1971, pp. 1621–6.
[38] Saaty, T. L. and J. Bram, *Nonlinear Mathematics*, Dover, New York, 1981.
[39] Shigley, J. E., *Mechanical Engineering Design*, 3rd ed., McGraw-Hill, New York, 1977.
[40] Schilling, R., "Effect of Unsprung Mass on Ride," *SAE Trans., Vol. 56*, 1948, pp. 220–6.
[41] Schlichting, H. (Translated by J. Kestin), *Boundary Layer Theory*, McGraw-Hill, New York, 1968.
[42] Smith, L. P., *Mathematical Models for Scientists and Engineers*, Prentice-Hall, Englewood Cliffs, NJ, 1953, pp. 36, 139–41.
[43] Smith, D. E., *History of Mathematics*, Vols. I & II, Dover, New York, 1958.
[44] Starfield, A. M., K. A. Smith, and A. L. Bleloch, *How to Model It, Problem Solving for the Computer Age*, McGraw-Hill, New York, 1990.
[45] Streeter, R. L. and J. K. Vennard, *Elementary Fluid Mechanics*, Wiley, New York, 1982, p. 534.
[46] Struik, D. J., *A Concise History of Mathematics*, 4th ed., Dover, Mineola, NY, 1987.
[47] Suh, N. P., *Principles of Design*, Oxford University Press, New York, 1990.
[48] Swokowski, E. W., *Calculus with Analytic Geometry*, Prindle, Weber, & Schmidt, Inc., MA, 1976, pp. 398–9.
[49] Taylor, B., *Methodus Incrementorum Directa & Inversa*, Pearsnianis, London, 1715.
[50] Thompson, D. E., "The Fluid Dynamics of Renal Blood Flow," Ph.D. dissertation, Dept. of Mechanical Engineering, Purdue University, Dissertation Abstracts International, Vol. 31, No. 70-18753, 1970.

References

[51] van der Pol, B., "The Non-linear Theory of Electric Oscillations," *Proc. Inst. Radio Eng.*, *Vol. 22*, p. 1051, 1934.

[52] Vanderplaats, G. N., *Numerical Optimization Techniques for Engineering Design with Applications*, McGraw-Hill, New York, 1984.

[53] Vollmer, E., *Leonardo DaVinci*, an Artabras Book, Reynal and Company in assoc. with William Morrow and Co., New York, 1956.

[54] Wylie, C. R., *Advanced Engineering Mathematics*, 4th ed, McGraw-Hill, New York, 1975.

[55] Zeid, I., *CAD/CAM Theory and Practice*, McGraw-Hill, New York, 1991.

Index

airplane, 250
airplane landing, 84
analytic methods
    Galerkin, 168
    harmonic balance, 163
    iteration, 155
    perturbation, 148
    power series, 157, 161
    secondary Galerkin, 171
automobile braking
    Liénard, 141
automobile engine, 231, 251
automobile handling, 44, 46
    analytic methods
        power series, 160
    direct integration, 99
    linearization, 95
    perturbation, 149
    variation of parameters, 102
automotive water pump, 262
autonomous equations, 137

Bagnall, J., 2
bang-bang ship rudder control, 145
bead on wire
    Galerkin, 173
    perturbation, 151
    Picard's method, 111
Bendixson's theorem, 206
bifurcation, 220
Blasius
    equation,
        2D boundary-layer problem, 45
        Galerkin solution, 174
    function, 46
blood flow, 242
blowgun, 252
Bogoliubov, N. N., 163
Bohr, Harald, 72
boundary-layer flow, 45
braking, bistable, 144

canoe, 253
carburetor, 229
catheter-balloon problem
    normalization, 56
chaotic
    pendulum, 217
chaotic system behavior, 217
climbers, 264
clutch, 253
column buckling
    Galerkin, 175
conservation equation
    energy, 40
    general, 35
    mass, 38, 45
    momentum, 39
continuity equation, 45

dc generator
    stability, 207
design, 1, 10
    analysis, 6
    definition, 4
    evaluation, 7
    ideas, 5
    identification, 4
    implementation, 7
    selection, 5
differential equations, 41
    linear, 42
    nonlinear, 42
    ordinary, 41
    partial, 41
Diophantus, 15
direct integration, 96
    automobile handling, 99
    gravity, 100
    spring-dashpot, 98
    thermal radiation, 97
dragster, 254
Duffing equation
    analytic methods, 159

Duffing equation (*cont.*)
  generalized, 161
  iteration, 156
  jump resonance, 159
  perturbation, 154
  power series, 159, 161
dynamics
  vehicle, 46

Earth cooling, 117, 248
Einstein, Albert, 25
elliptic integrals, 103, 116
  pendulum, 106
  suspended mass, 104
energy recovery, 255
equivalent linear system, 166, 182
errors
  least square, 75
  measures, 73
  Taylor series, 94
  weighted square, 76
Euler equation, 43
Euler methods, 122
  higher-order systems, 122
  modified, 121
  spring-mass, 123
Euler, Leonhard, 120
exact solution, 85, 114, 115, 183

finger model, 245
first-order system, 51
first term neglected
  Taylor-series errors, 95
flow regulator
  linearization, 90
fluid mechanics, 45
flyball governor
  linearization, 91
  stability, 186
forearm, 249

Galerkin, 182, 183, 263
  analytic methods, 168
  bead on wire, 173
  Blasius problem, 174
  column buckling, 175
  heat conduction, 177
  heat radiation, 263
  helicopter model, 169
  linear damped muscle, 170
  nonlinear spring-mass, 179
Galilei, Galileo, 44
Gauss, Carl Friedrich, 26
gedanken, 14
Gill's constants, 127

glass house, 255
graphical methods
  Liénard, 139
  isoclines
    first-order systems, 133
    second-order systems, 137
  Pell, 142
  phase-plane analysis, 135
    independent variable recovery, 136
graphical solutions, 145, 147, 251
  Lienard's, 144, 147
gravitational attraction, 86
gravity, 12
  direct integration, 100
  linearization, 89, 96
Green's theorem, 204
growing phenomenon, 116, 145

harmonic balance, 182
  analytic methods, 163
  spring-mass-dashpot, 165
  van der Pol oscillator, 165, 167
harmonic oscillator problem
  Runge-Kutta, 127
heat conduction
  Galerkin, 177
heat transfer, 43, 86, 263
  radiation, 96
helicopter, 257
  model
    Galerkin, 169
history, 263
hydroelectric generating system, 233

identity matrix, 190
independent variable recovery, 136, 147
indicial notation, 25
infinite-domain transformation, 60
isoclines, 134, 144, 145, 147
  graphical methods, 133
  second-order systems, 137
  spring-dashpot, 134
iteration, 181
  analytic methods, 155
  Duffing equation, 156

jump resonance, 159, 161

Koberg, D., 2
Krylov, A. N., 163
Kutta, Wilhelm Martin, 123

L'Hôpital's rule, 197
Lagrange modeling techniques, 22
Lagrange, Joseph-Louis, 23

Index

Legendre, Adrien-Marie, 75, 107
Leibnitz' rule, 38
Leibnitz, Gottfried Wilheim, 38
Leonardo da Vinci, 1
Liénard, Alfred, 139
Liénard
    automobile braking, 141
    graphical methods, 139
    van der Pol oscillator, 141
lightbulb, 226, 256
linear damped muscle
    Galerkin, 170
linearization, 85–7, 182
    automobile handling, 95
    flow regulator, 90
    flyball governor, 91
    multivariable, 90
    trigonometric function, 89
Lyapunov
    first theorem, 212
    function, 211
    second method, 210
    second theorem, 215
Lyapunov, A. M., 210

mass on cubic spring, 115, 116, 131, 181, 182
mass on square-law spring, 181, 182
mass on square-law spring, damper, 116, 183
math model, 10
matrix, zero, 190
McPhate, A. J., 6
mechanics, 44
metering pin damper, 116, 182, 183
method of perturbation, 181
model
    airplane, 250
    automobile engine, 231, 251
    automotive water pump, 262
    "best," 72
    blood flow in kidney, 242
    blowgun, 252
    canoe, 253
    carburetor, 229
    catheter-balloon, 56
    chain falling, 253
    class, 253
    climbers, 264
    clutch, 253
    dragster, 254
    Earth cooling, 248
    economy, 223
    eggs in a box, 18
    energy recovery, 255
    finger, 245
    forearm, 249
    garage door, 225
    gas tank dilution, 258
    glass house, 255
    helicopter, 257
    hydroelectric generating system, 233
    lightbulb, 226
    mass on wire, 114
    meter pin damper, 116
    muscle motor, 248, 249
    ping-pong launcher, 258
    population, 223
    RC-circuit, 259
    renal, 242
    rocket, 260
    sail, 258
    shock absorber, 236
    temperature-dependent circuit, 51, 125
    thermal and electrical inertia, 132
    tire growth, 260
    transmission, 262
    two-body gravitational attraction, 86
    whale, 20, 262
modeling, 10
modeling techniques
    Lagrange, 22
    normalization of equations, 47
multistep
    numerical methods, 128
muscle motor design, 248, 249

Navier, Louis M. H., 45
Navier-Stokes, 45, 64
Newton, Isaac, 12
no-crossing property, 218
nonlinear spring-mass
    Galerkin, 179
    secondary Galerkin, 172
normalization, 84, 131
    catheter-balloon problem, 56
    complex systems, 59
    infinite, semi-infinite domain transformation, 60
    modeling technique, 47
    pipe draining a reservoir, 58
    spring-mass-dashpot system, 49
numerical methods, 118
    Euler, 120
    Euler modified, 121
    multistep, 128
    Runge-Kutta, 123
        spring-dashpot, 125
    Taylor series, 118
numerical solution, 131
Nyquist, 186

optimization
  gradient method, 78
  nongradient method, 78
orthogonal functions, 76

Pell
  equation, 142
  graphical method, 142
  pendulum problem, 143
Pell, John, 142
penalty functions, 81
pendulum, 44
  chaotic, 217
  elliptic integrals, 106
  Pell method, 143
perturbation
  analytic methods, 148
  automobile handling, 149
  bead on wire, 151
  Duffing equation, 154
phase-plane, 218, 221, 223, 224
  conservative systems, 219
  dissipative systems, 219
  no-crossing property, 218
  Poincaré sections, 219
phase-plane analysis
  graphical methods, 135
  independent-variable recovery, 136
  spring-dashpot, 136
phase space, 218
Picard solution, 115–117
Picard's method, 109
  bead on wire, 111
  spring-mass-dashpot, 110
  uniqueness, 110
Picard, Charles Emil, 109
Poincaré, H., 189, 217
  sections, 219
Porsche/Mercedes Benz race car, 260
power series, 108
  analytic methods, 157
  automobile handling, 160
  reversion method, 112
  spring-damper, 108
power-series solution, 116
Principle of Minimum Astonishment, 181
properties of nonlinear systems, 13

Rayleigh equation, 147
reversion method
  power series, 112
Robinson's oscillator, 144
rocket, 260
rocket car, 39

rod kinematics, 264
Routh-Hurwitz, 186
Runge, Carl David Tolmé, 123
Runge-Kutta, 123
  harmonic oscillator problem, 127
  numerical methods, 123
  system of equations, 126

sail, 258
Schlichting, H., 45
secondary Galerkin
  analytic methods, 171
  nonlinear spring-mass, 172
secular term, 152, 156
separation of variables, 64
separatrix, 221
Shigley, J. E., 2
singular-point analysis, 221–225
  stability, 187
slowly varying parameters, 182
spring-damper
  power series, 108
spring-dashpot
  direct integration, 98
  exponential transformation, 62
  isoclines, 134
  numerical methods
    Runge-Kutta, 125
  phase-plane analysis, 136
    independent-variable recovery, 136
  stability, 184
  Taylor series, 119
spring-mass
  analytic methods
    power series, 158
  Euler method, 123
spring-mass-dashpot
  harmonic balance, 165
  Picard method, 110
stability, 147, 184, 221–5
  flyball governor, 186
  Lyapunov method, 223
  singular-point analysis, 187
  spring-dashpot, 184
Stokes' theorem, 26
Stokes, George Gabriel, 26, 45
subharmonics, 161, 162
suspended mass
  elliptic integrals, 104
symbolic manipulation, 85, 86, 115, 131, 181, 182, 252, 256, 263
system behavior
  nonlinear, 13

Index

Taylor series, 88
  errors, 94
    classical, 94
    first term neglected, 95
  numerical methods, 118
  spring-dashpot, 119
Taylor, Brook, 88
temperature-dependent circuit, 84
tensor
  algebra, 30
  alternating, 32
  antisymmetric, 31
  definition, 28
  derivative, 33
  invariant, 31
  isotropic, 33
  notation, 25, 29
  symmetric, 31
thermal and electrical inertia, 85, 132
thermal radiation, 86, 96
  direct integration, 97
thermodynamics, 43
tire model, 260
transformation, 81
  exponential, 61
  generalized, 62
  linear, 60
  mapping, 61
  normalization, 47, 60
  similarity, 45, 63
  vector, 27
transmission automobile, 262
trigonometric function
  linearization, 89

van der Pol, B., 165
van der Pol oscillator, 206
  harmonic balance, 165, 167
  Liénard, 141
variable mapping, 81
variation of parameters, 101
  automobile handling, 102
vector, 27
  transformation, 27
vibrating beam, 23
viscosimeter, 265
Volterra's equations, 225

whale model, 20, 262

Yûnis, Ibn, the Younger, 44

UMIST Library and Information Service